Innovative Approaches in Asphalt Binder Modification and Performance

Innovative Approaches in Asphalt Binder Modification and Performance

Guest Editors

Yong Wen

Xiaokang Zhao

MDPI • Basel • Beijing • Wuhan • Barcelona • Belgrade • Novi Sad • Cluj • Manchester

Guest Editors
Yong Wen
School of Highway
Chang'an University
Xi'an
China

Xiaokang Zhao
School of Highway
Chang'an University
Xi'an
China

Editorial Office
MDPI AG
Grosspeteranlage 5
4052 Basel, Switzerland

This is a reprint of the Special Issue, published open access by the journal *Materials* (ISSN 1996-1944), freely accessible at: https://www.mdpi.com/journal/materials/special_issues/XPD113EN5G.

For citation purposes, cite each article independently as indicated on the article page online and as indicated below:

Lastname, A.A.; Lastname, B.B. Article Title. *Journal Name* **Year**, *Volume Number*, Page Range.

ISBN 978-3-7258-7394-4 (Hbk)
ISBN 978-3-7258-7395-1 (PDF)
https://doi.org/10.3390/books978-3-7258-7395-1

Contents

About the Editors . . . . **vii**

Preface . . . . **ix**

Oswaldo Guerrero-Bustamante, Amparo Guillen, Fernando Moreno-Navarro, M. C. Rubio-Gámez and Miguel Sol-Sánchez
Suitable Granular Road Base from Reclaimed Asphalt Pavement
Reprinted from: *Materials* **2025**, *18*, 854, https://doi.org/10.3390/ma18040854 . . . . **1**

Jinchao Yue, Jiahao Fu, Yihan Wang, Yang Ming, Guoqi Tang and Ruixia Li
Effects of Complex Environmental Conditions on Fatigue Self-Healing Properties of Fast-Melting SBS-Modified Asphalt
Reprinted from: *Materials* **2025**, *18*, 2157, https://doi.org/10.3390/ma18092157 . . . . **21**

Marek Pszczola and Bohdan Dolzycki
Investigation of Rutting and Aging Performance of Gap-Graded Rubberized Asphalt Mixtures
Reprinted from: *Materials* **2025**, *18*, 2263, https://doi.org/10.3390/ma18102263 . . . . **39**

Zhen Shen, Shikun Wang, Zhe Hu and Xiaokang Zhao
Research on Crack Resistance of Foamed Rubber Asphalt Cold Recycled Mixtures Based on Semi-Circular Bending Test
Reprinted from: *Materials* **2025**, *18*, 2684, https://doi.org/10.3390/ma18122684 . . . . **51**

Pei Li, Rongyi Ji, Chenlong Zhang, Jinghan Xu, Mulian Zheng and Xinghan Song
A Study on the Road Performance of the Self-Healing Microcapsule for Asphalt Pavement
Reprinted from: *Materials* **2025**, *18*, 3483, https://doi.org/10.3390/ma18153483 . . . . **68**

Hao Wu, Daoan Yu, Wentao Wang, Chuanqi Yan, Rui Xiao, Rong Chen, et al.
A Study on the Characterization of Asphalt Plant Reclaimed Powder Using Fourier Transform Infrared Spectroscopy
Reprinted from: *Materials* **2025**, *18*, 3660, https://doi.org/10.3390/ma18153660 . . . . **89**

Yan Zhang, Jiyang Li and Yiren Sun
Evaluation and Control of Variability in RAP Properties Through Refined Fractionation Processing Methods
Reprinted from: *Materials* **2025**, *18*, 4944, https://doi.org/10.3390/ma18214944 . . . . **108**

Maciej Krasowski, Przemysław Buczyński, Grzegorz Mazurek and Matúš Kozel
New Type of Asphalt Concrete with Bitumen Emulsion in Terms of Type and Quantity of Binder
Reprinted from: *Materials* **2025**, *18*, 5437, https://doi.org/10.3390/ma18235437 . . . . **125**

About the Editors

Yong Wen

Yong Wen is an Associate Professor at Chang'an University, China. His research is primarily dedicated to road engineering, with a specific focus on the aging behavior and mechanisms of asphalt, the development of green and environmentally friendly paving materials, asphalt pavement durability, and structural safety early warning systems. He has presided over and participated in numerous significant research projects, including sub-tasks of the National Key R&D Program of China, the National Natural Science Foundation of China, and the China Postdoctoral Science Foundation. Additionally, he has led projects funded by the Shaanxi Provincial Natural Science Basic Research Program.

Xiaokang Zhao

Xiaokang Zhao is an Associate Professor and Postgraduate Supervisor. He holds a Ph.D. in Engineering. His primary research focuses on road and airport pavement engineering, with specific interests in multi-scale characterization and the simulation of pavement materials, the development of durable pavement materials, intelligent pavement systems with perception and monitoring, and pavement performance detection and maintenance technologies. He has published over 50 academic papers and holds more than 20 authorized patents. Dr. Zhao has presided over two projects supported by the National Natural Science Foundation of China, eight science and technology projects at provincial/ministerial levels, and four civil aviation science and technology research projects.

Preface

Asphalt binders are fundamental to the structural integrity of our transportation networks, yet they face continuous challenges from environmental stressors and heavy traffic demands. The motivation behind this Special Issue is to consolidate cutting-edge research that overcomes the limitations of traditional binders through advanced modification techniques.

This Special Issue encompasses varied innovative approaches, including the application of novel modifiers such as polymers, nanomaterials, and bio-based agents. Our objective is to highlight technologies that significantly improve the durability, rheology, and aging resistance of asphalt materials. Furthermore, this Special Issue places a strong emphasis on sustainability, presenting research on recycled materials and environmentally friendly additives that reduce the ecological footprint of pavement construction without compromising performance.

This Special Issue is targeted at pavement engineers, materials scientists, and researchers in the field of civil engineering. We trust that the findings presented here will contribute to the optimization of binder formulations for diverse climates and conditions, ultimately advancing the sustainability and resilience of global infrastructure.

Yong Wen and Xiaokang Zhao
Guest Editors

materials

MDPI

Article

Suitable Granular Road Base from Reclaimed Asphalt Pavement

Oswaldo Guerrero-Bustamante [1,2], Amparo Guillen [1], Fernando Moreno-Navarro [1], M. C. Rubio-Gámez [1] and Miguel Sol-Sánchez [1,*]

[1] Construction Engineering Laboratory, University of Granada (LabIC.UGR), 18071 Granada, Spain; oguerrer5@cuc.edu.co (O.G.-B.); amguillen@ugr.es (A.G.); fmoreno@ugr.es (F.M.-N.); mcrubio@ugr.es (M.C.R.-G.)

[2] Department of Civil & Environmental, Universidad de la Costa, Barranquilla 080003, Colombia

* Correspondence: msol@ugr.es

Abstract: The granular bases commonly used in the construction of road infrastructure projects often require a high consumption of raw materials. The potential utilization of recycled materials, specifically Reclaimed Asphalt Pavement (RAP) derived from road asphalt pavement demolition, emerges as a promising sustainable advantage for infrastructure projects, considering its potential environmental and cost benefits in other layers of the structure. In this context, this research proposes a feasibility study on the use of RAP as a granular base layer, supported by an advanced laboratory analysis that includes a range of tests simulating the in-service conditions as well as a full-scale demonstration of the material behavior under static and dynamic loads. Various design variables, such as different gradations and binder content, are considered. The results demonstrate that, despite having discontinuous gradation and smaller aggregate sizes than those commonly applied in natural base layers, the evaluated recycled materials exhibit a higher load-bearing capacity and resistance to permanent deformation than the reference materials commonly used as granular bases. Notable enhancements of up to 30% in elastic modulus, coupled with reductions of around 20% in permanent deformations, have been documented using the asphalt cement potential in the old pavement.

Keywords: granular base; Reclaimed Asphalt Pavement (RAP); sustainable materials; road infrastructure

1. Introduction

The construction and maintenance of road infrastructures represent a constant challenge in the field of civil engineering and sustainable development. In this context, the search for alternatives that not only meet technical performance standards but also address environmental and economic concerns has become a strategic priority [1,2]. Reutilizing waste materials has emerged as a viable alternative for road maintenance, rehabilitation, and even in the construction of new pavements [3]. One of the most promising approaches is the use of Reclaimed Asphalt Pavement (RAP), as a key component in the structural layers of pavement [4].

The global demand for aggregates in the construction sector is estimated to be between 20 and 40 billion tons annually. This demand is projected to grow at an annual rate of about 6.8% until 2031 [5,6], positioning the construction sector as one of the most environmentally impactful industries [7]. The sector is associated with 40% of global greenhouse gas emissions and accounts for 30% of worldwide water consumption [8–10]. The generation of demolition waste worldwide remains a significant problem. In Australia, approximately

1.2 million tons of RAP are produced annually [11], while in the United States, construction waste production reaches 30 million tons [12]. In many cases, only 40% of this waste is reused in the construction sector [13].

While RAP is primarily reused in asphalt mixtures [14–16], its incorporation is often limited to approximately 20–30% due to logistical constraints and regulatory restrictions [17]. This restricted use highlights the need to explore alternative applications for RAP to maximize its potential in infrastructure projects. One such application is its use in granular base layers, which play a fundamental role in the stability and durability of pavement structures. Traditionally, these layers have been constructed using natural or virgin aggregates, requiring extensive resource extraction and raising environmental concerns. In this context, RAP emerges as a viable and sustainable alternative.

A report from the Virginia Center for Transportation Innovation and Research [18] demonstrated the technical feasibility of using RAP in unbound base layers, a practice already adopted by numerous agencies in the United States. However, there is a lack of technical literature compiling the current field performance of these applications, highlighting the need for further research into the mechanical behavior and durability of RAP-based granular layers under real-world service conditions.

RAP has appropriate characteristics that make it appealing in terms of sustainability [4,19]. Its composition, enriched with residual asphalt, can contribute to the load-bearing capacity and strength of the material [20]. However, some studies have indicated reductions in CBR strength [21] and compacted density of the materials when RAP is included [21,22]. Conversely, increases in the resilient modulus have been observed [22]. Previous experience has shown the effective utilization of reclaimed asphalt pavement (RAP) in granular layer applications. Recent research highlights the testing of mixes incorporating 50% and 75% RAP alongside fresh granular materials and recycled concrete aggregate to evaluate their suitability for granular base layers [3,23]. The incorporation of RAP in the construction of granular base and sub-base layers not only imparts structural advantages but also leads to cost savings and environmental benefits by diminishing reliance on natural aggregates and promoting sustainability [24–26]. Numerous studies have indicated that granular layers incorporating RAP can deliver strength equal to or surpassing that of traditional base and sub-base layers, thereby contributing to economically viable and sustainable road construction practices [27].

Despite its potential, the mechanical properties of RAP-based granular layers remain insufficiently studied [3,28], particularly under realistic service conditions. The heterogeneity of RAP due to variations in aggregate composition, asphalt content, and processing methods necessitates further investigation. Advanced studies, including full-scale evaluations, are required to assess the mechanical behavior of RAP-based layers comprehensively and ensure their reliable implementation in infrastructure projects.

In general, to fully power these advantages, it is essential to thoroughly understand its mechanical behavior, its interaction with other pavement layers, and its response to specific design variables, considering that this material from the crushing or milling of asphalt pavements is highly heterogeneous and possesses unique mechanical properties. Thus, the primary purpose of this research is to explore the technical feasibility of using RAP from the demolition of deteriorated pavements as a granular base layer in new pavement structures under expected service conditions, supported by an advanced laboratory study that includes the incorporation of RAP with different physical properties. The research includes physical-mechanical characterization assessing different variables while carrying out full-scale testing box evaluation of the material mechanical behavior in terms of load-bearing capacity and resistance to permanent deformations, considering the current limited number of studies thoroughly investigating this application under real service conditions.

2. Materials and Methods

2.1. Materials

In this study, four materials were employed: one natural reference aggregate and three RAP samples with different characteristics to analyze the impact of their property variability (e.g., source and processing) on the feasibility assessment. The natural reference aggregate corresponds to a type of crushed stone of limestone origin, commonly used as a granular base in road infrastructure in Spain. This material exhibits a continuous particle size distribution with particle fractions in the range of 0.0/32.0 mm. Moreover, the granular material exhibited suitable physical and mechanical properties for its application as a granular base, adhering to Spanish standards [29].

The three types of RAP (referred to as RAP 1, RAP 2, and RAP 3) were collected from different roadways in Andalusia (Spain), then processed at a plant, crushed, and classified into 0/22 mm fractions to ensure uniformity and obtain a more representative sample for use in this study. These materials exhibit different particle size distributions and binder content, which makes them appropriate for use in this study, which analyzes the influence of RAP variability. The main properties are listed in Table 1. It is noted that the recycled materials have varying amounts of residual binder content, ranging from 3.0% to 4.0% of the total weight.

Table 1. Materials' main properties.

Properties	Standard	Crushed Stone	RAP 1	RAP 2	RAP 3
Flakiness index (%)	UNE EN 933-3:2023 [30]	18	13	6	11
Percentage of fractured face (%)	UNE-EN 933-5:2023 [31]	100	100	100	100
Sand equivalent (%)	UNE-EN 933-8:2015 [32]	55	78	81	69
Resistance to fragmentation (L.A.) (4/8 mm) (%)	UNE EN 1097-2:2021 [33]	23.9	20.7	15.3	20.8
Resistance to fragmentation (L.A.) (8/11.2 mm) (%)	UNE EN 1097-2:2021 [33]	20.8	17.1	12.8	18.7
Density (Mg/m^3)	UNE-EN 1097-6 [34]	2.81	2.57	2.57	2.56
Coefficient of uniformity (Cu)	UNE-EN 933-1:2012 [35]	45	8	10	11
Curvature coefficient (Cc)	UNE-EN 933-1:2012 [35]	3.1	3.1	3.1	1.4
Modified Proctor Opt. Moisture content (%)/Max Dry Density. (g/cm^3)	UNE 103501:1994 [36]	5.0/2.37	5.5/2.25	5.5/2.32	5.5/2.33
Residual binder content (% Total weight)	Recovery of binder by Evaporation UNE-EN 12697-1:2013 [37]	-	3.9	3.0	4.0
Binder Penetration (25 °C) (0.01 mm)	UNE-EN 1426:2015 [38]	-	16.0	4.0	8.0
Softening Point (°C)	UNE-EN1427:2015 [39]	-	71	129	85

Similarly, it can be observed in Figure 1 that there are RAP materials with different particle size distributions; RAP 1 consists of fine particles within the particle size range of 0/16.0 mm, whereas RAP 2 is characterized by coarser particles, falling within the range of 0.0 to 20.0 mm. Lastly, RAP 3 is categorized as an intermediate material, with particle sizes ranging from 0.0 to 20.0 mm.

2.2. Testing Plan

The testing plan conducted in this research is presented in Table 2; it consists of five main stages: the first one consists of the (i) study of the mechanical behavior and parametric analysis of designing variables for the RAP base layer in order to study the effect of the RAP type and RAP replacement on natural aggregates in different proportions; subsequently, for the materials exhibiting optimal performance in this initial stage, (ii) bearing capacity using a 300 mm load plate, and (iii) resistance to permanent deformation tests were conducted in a 35 cm × 35 cm test box, aiming to obtain preliminary information on the mechanical behavior of these materials. In the same direction, (iv) the ability to protect the lower layers was tested [40], and (v) a full-scale laboratory testing box to analyze the behavior under real operating conditions.

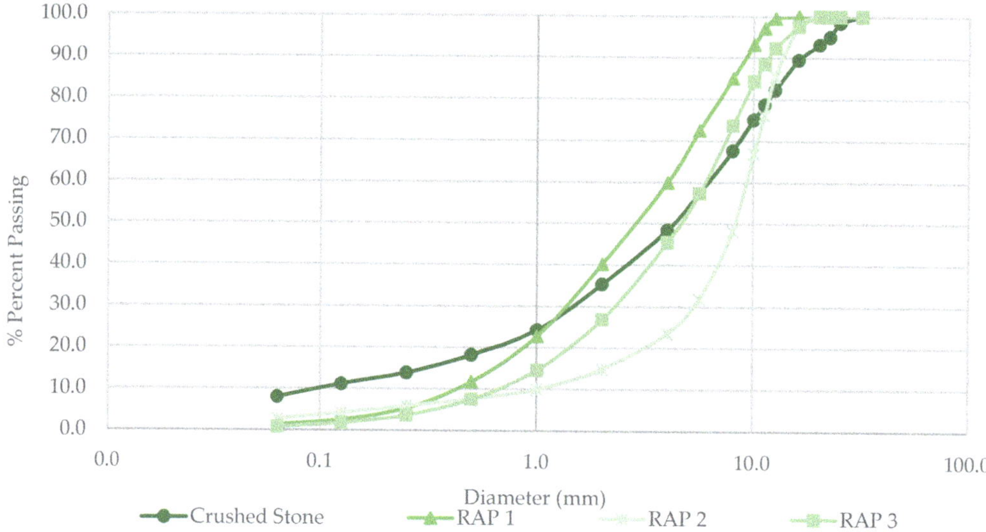

Figure 1. Particle size distribution.

Table 2. Testing plan.

Stage	Properties	Materials	Test
Optimization Study of the Use of RAP as a Granular Layer	Impact of RAP Type Impact of Combining RAP with Reference Materials	Crushed Stone RAP 1 RAP 2 RAP 3	Indirect Tensile Strength Compressive Strength
Bearing Capacity	Elastic Modulus	Crushed Stone Opt. RAP	300 mm Static Plate Bearing Test 300 mm Dynamic Punching Test
Permanent Deformation	Total Rut Depth		Wheel Tracking Test
Permeability	Vertical Permeability (K)		Permeameter
Full Scale Laboratory Testing Box	Bearing CapacityPermanent Deformation		600 mm Static Plate Bearing Test 600 mm Dynamic Punching Test Permanent Deformation

To study the mechanical behavior of the recycled materials and the crushed stone (reference), this research examines the effects of different design variables in the use of RAP for the base layer, analyzing their influence on crucial mechanical properties as follows:

- Influence of RAP type: To achieve this goal, cylindrical specimens were manufactured and compacted with the optimal moisture content determined in the characterization phase. For specimens containing recycled materials, re-compaction was performed 24 h later at a temperature of 60 °C, considering the presence of a thin layer of adhered bitumen that enhances the mechanical behavior of the material [41] and that this temperature can be reached on-site [42–44]. A direct comparison was made between the reference materials and the recycled materials.

- Impact of combining RAP with reference materials: In this part of the study, the effect of replacing RAP in the crushed stone on mechanical behavior was evaluated with the aim of assessing the viability of partial replacement of virgin material with recycled material, in this case, in a 25/75, 50/50 and 75/25 recycled/virgin material ratio based on previous studies [45–47].

In this case, the tests proposed to evaluate the mechanical behavior are simple compression strength and indirect tensile strength; they were selected based on their suitability for assessing granular materials. The simple compression test is particularly appropriate for unbound granular materials, providing insights into their load-bearing capacity under compressive loads. Meanwhile, the indirect tensile strength test was chosen to investigate whether the presence of adhered bitumen in RAP materials contributes to tensile resistance, potentially enhancing the cohesion of the mixture [48].

For the simple compression test, cylindrical specimens with a diameter of 100 mm and a height of 200 mm were used, using granular materials (reference and RAPs) without any binder and compacted with the gyratory compactor according to standard UNE 103400:1993 [49]. The indirect tensile strength tests were conducted on cylindrical samples with a diameter of 100 mm and a height of 60 mm based on standard UNE-EN 12697-23:2018 [50]. To ensure the reliability of the results, three specimens were prepared for each test. As mentioned before, the cylindrical specimens were manufactured with the optimal moisture content found in the material characterization phase. Additionally, for specimens that included recycled materials (RAPs), were compacted again 24 h later at a temperature of 60 °C, aiming to take advantage of the agglomeration capacity provided by the residual bitumen.

For the recycled material with better performance (Opt. RAP) and the reference material, a load-bearing test was conducted using the 300 mm static and dynamic plate load test according to standards UNE 103808:2006 [51] and UNE 103807-2 [52], respectively. To ensure the reliability of the results, three test boxes were prepared for each condition. For the static load test, two loading cycles were performed, each with six steps with approximately equal increments (85 kPa, 200 kPa, 300 kPa, 400 kPa, 500 kPa, and 600 kPa). The dynamic plate load test, on the other hand, was conducted by applying loading cycles with an amplitude of 420 kPa at a frequency of 5 Hz. Both tests allow for determining the vertical modulus of elasticity under static load (E_V) and dynamic load (E_{DYN5HZ}) for each of the studied materials.

On these same materials, permanent deformation or rutting tests were conducted using the Wheel Tracking device using two specimens for each condition according to the UNE-EN 12697-22 standard [53]. In this case, the material's susceptibility to deformation is evaluated by measuring the rut depth produced by repeated passes at a fixed temperature of 25 °C. A wheel is subjected to a standardized load of 700 N with a total travel of 230 mm at an approximate frequency of 26 cycles per minute.

A constant head permeability test adapted from standard UNE-EN 103403:1999 [54] was conducted using a 300 mm × 300 mm box with a height of 500 mm, as can be seen in Figure 2. For each material, three tests were performed. The coefficient of permeability was calculated for the studied materials, defined as the water flow through a unit cross-sectional area under a unit hydraulic gradient in conditions of laminar flow using Darcy's law. This parameter is valuable for assessing the material's ability to waterproof the rest of the substructure.

Finally, with the aim of simulating real operating and confinement conditions, a test box of 100 cm × 100 cm was constructed. The samples under study were layered with a thickness of 150 mm, compacted at the optimal moisture content, and in the case of RAP, re-compacted 24 h later at 60 °C using ultraviolet lights. A BBTM-11B asphalt mixture

(UNE-EN 13108-2 [55]), produced with PMB 45/80-60 bitumen, with a thickness of 50 mm, was placed on the reference materials, as shown in Figure 3.

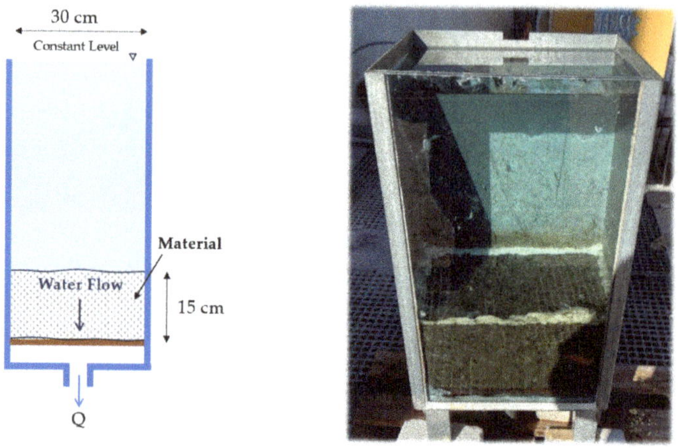

(**a**) Test Scheme (**b**) Testing Box

Figure 2. Permeability test setup.

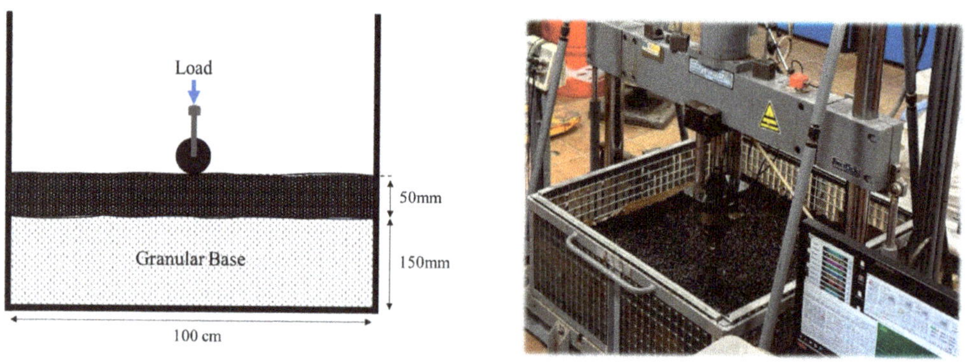

(**a**) Scheme (**b**) Testing Box

Figure 3. Full-scale testing box.

On the samples previously prepared in the test box were conducted load-bearing tests, using a 600 mm loading plate and under static and dynamic loads according to the standards UNE 103808:2006 [51] and UNE 103807-2 [52], respectively, using a 100 kN capacity loading frame, as described by Sol-Sánchez et al. [56]. For the first case, 2 cycles of load and unload were applied, with load levels of 35 kPa, 70 kPa, 105 kPa, 150 kPa, 175 kPa, 210 kPa, and 245 kPa. In the dynamic load test, load cycles of 180 kPa with a frequency of 5 Hz were used. Finally, permanent deformation tests were conducted by applying 110,000 load cycles with an amplitude of 700 kPa and a frequency of 5 Hz, using a wheel with a diameter of 20 cm and a thickness of 40 mm. Figure 4 shows some of the conducted tests.

(a) Wheel Tracking Test

(b) Particle Size Distribution

(c) Bearing Capacity

(d) Permanent Deformation

Figure 4. Setup of some of the conducted tests.

3. Results and Discussion

3.1. Optimization Study of the Use of RAP as a Granular Layer

3.1.1. Influence of RAP Type

Figure 5 shows the results of (a) simple compression and (b) indirect tensile strength tests conducted in this study, comparing the strength of different types of RAP (compacted at 60 °C to provide the potential bonding benefits from residual asphalt binder in the recycled aggregates) in reference to natural crushed stones. The results indicate that the average compression strength of recycled materials was higher than that of the reference material, showing an average improvement of 65%. This aligns with findings reported by Alam et al. [57], who observed the competitive mechanical performance of RAP materials compacted at optimal moisture and ambient temperature under similar conditions, highlighting the potential of RAP to achieve comparable or superior performance to natural aggregates. In this case, RAP 1 and RAP 3, which have a more continuous gradation, exhibited the best performance, which could be related to a stringer mineral skeleton in comparison to RAP 2.

More significant improvements were observed in indirect tensile strength, with recycled materials outperforming reference materials by an average of 126%. This notable increase highlights the critical role of the bonding capacity provided by the residual bitumen in RAP, which contributes significantly to resistance under tensile stresses and increases the bearing capacity [19]. Similarly, RAP 1 and RAP 3 demonstrated superior performance, while RAP 2, with a more discontinuous gradation and lower asphalt content, exhibited inferior behavior in both compression and indirect tensile stresses. These findings indicate that the type of RAP plays a crucial role in enhancing the performance of sustainable granular layers, emphasizing the importance of carefully selecting RAP materials based on their gradation and asphalt content.

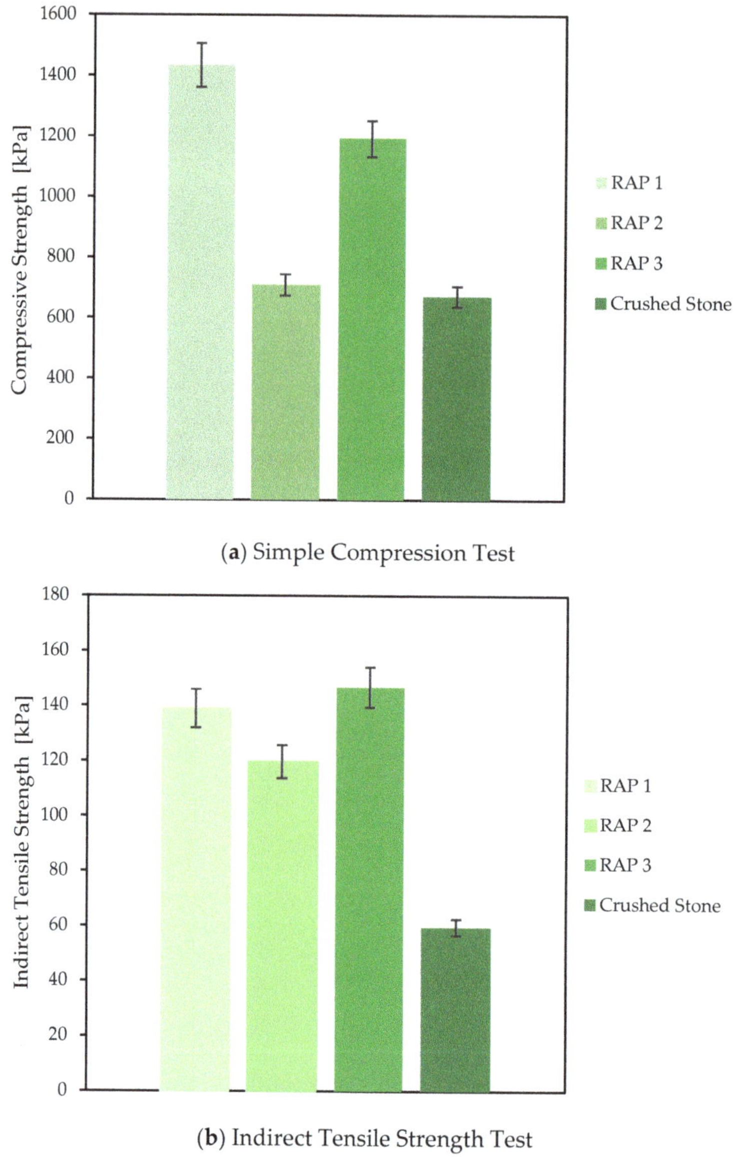

Figure 5. Indirect tensile strength and compression test results.

In light of the potential impact of RAP characteristics on its mechanical performance, correlations were performed between the physical properties of the materials (Table 1) and the measured compressive and tensile strengths in this part of the study. Figure 6 presents the coefficients of determination (R^2) for each parameter. The farther away the points are from the center of the graph, the stronger the correlation. Both measured mechanical parameters show a significant relationship ($R^2 > 60\%$) with the maximum aggregate size and filler content (% material passing the 0.0063 mm sieve).

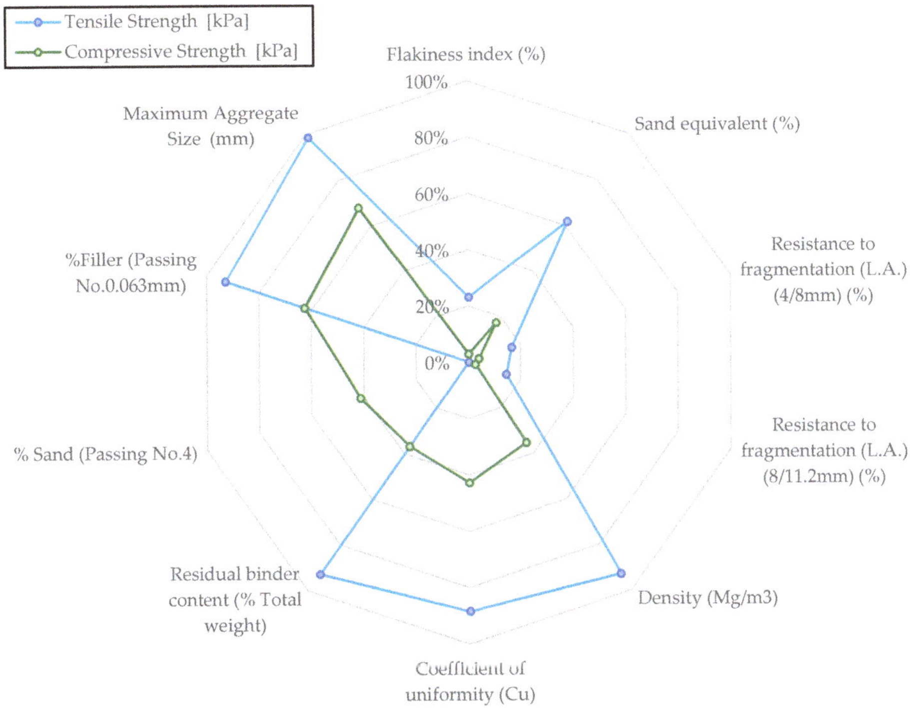

Figure 6. Laboratory materials correlations.

The indirect tensile strength appears to correlate well with density, particle size distribution measured with the coefficient of uniformity (Cu), and the residual asphalt content of the recycled material, contributing to the overall adhesiveness and mechanical integrity of the materials [58,59]. On the other hand, the compressive strength is closely related to the maximum density obtained from the Modified Proctor test. This suggests that while aspects such as gradation can be optimized during material selection and mix design, other inherent properties of RAP, such as binder content and fine aggregate quality, also play a critical role and must be carefully considered to ensure structural performance.

3.1.2. Impact of Combining RAP with Reference Materials

To analyze the effect of combining RAP with natural aggregates, as examined in previous studies [60,61]. Figure 7 shows the results of indirect tensile strength and simple compression tests for proposed replacements of 25%, 50%, and 75%, along with cases of 100% crushed stone and 100% RAP type 3 (best performed RAP). For this analysis, the results of combining RAP 3 are shown, as it exhibited a more continuous gradation, which is a key factor according to previous findings. The findings suggest that, in all cases, the mechanical performance of the crushed stone is enhanced with the incorporation of RAP, reaching conditions very similar to those of RAP (>80% of indirect tensile strength and simple compression) with replacements starting from 50% by weight of the recycled material.

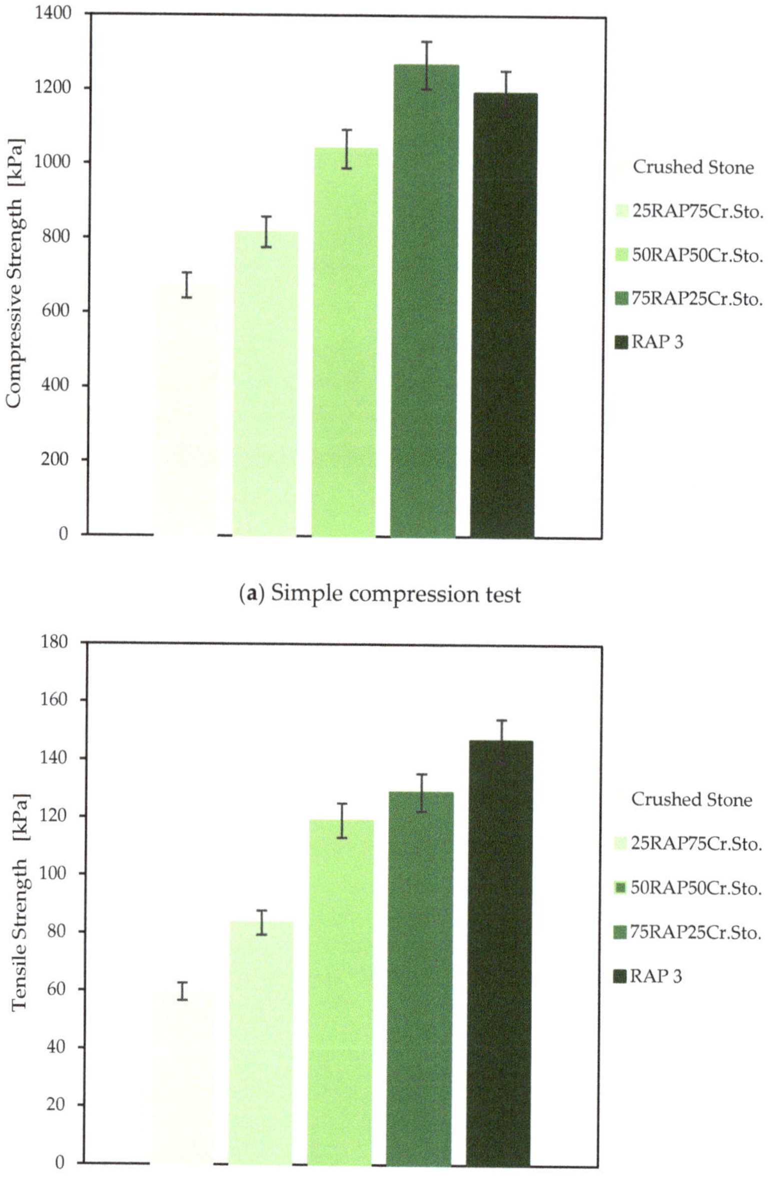

Figure 7. Impact of combining RAP with reference materials on mechanical behavior.

However, it is important to note that some studies limit the use of RAP to proportions around 40–50% [62], finding reductions in properties such as CBR strength when proportions exceed 60% [3]. These limitations are primarily associated with excessive deformations observed at higher replacement levels [63]. Nevertheless, the contrast with the results reported in this study may be attributed to the fact that the material was compacted at its optimal moisture content and, subsequently, re-compacted at high ambient temperatures. Under these conditions, the residual bitumen in RAP can start to enhance cohesion, improving mechanical performance [19]. This aspect is crucial for future applications, highlighting the importance of compaction temperature as a key factor influencing the performance of RAP-based granular layers.

Moreover, Moreover, while high RAP content may raise concerns about compliance with standard engineering specifications, particularly regarding particle size distribution and shape indices, partial replacement strategies offer a viable solution. By adjusting the proportion of RAP in the mix, it is possible to balance mechanical performance while maintaining compliance with engineering standards. Previous studies [3,64] suggest that partial replacements of RAP can influence physical properties, such as increased resistance to LA abrasion, decreased absorptions, and maximum dry densities. These findings reinforce the potential of partial RAP incorporation as a strategy to mitigate the drawbacks associated with high RAP content, while still maximizing the sustainability benefits of recycled materials.

3.2. Bearing Capacity

Considering the previous outcomes, the subsequent analysis explores the mechanical behavior to its application as a base layer, with a particular emphasis on RAP 3 (considered the optimum RAP in this study). Figure 8 summarizes the results of the static plate load test, similar to that reported by other authors [65,66]. The results indicate that the use of RAP resulted in values comparable to those recorded by the reference material. In addition, it is crucial to note that the total deformation after the two loading cycles was lower for the recycled material compared to the reference material, which resulted in a lower modulus ratio (E_{V2}/E_{V1}), indicating a layer with superior mechanical performance in terms of load bearing capacity. Therefore, these results suggest that layers with high RAP content do not necessarily lead to excessive deformations and reduced bearing capacity, as observed in previous studies [19,67]. When the residual bitumen is effectively utilized, RAP can provide significant mechanical advantages, enhancing cohesion and overall structural performance.

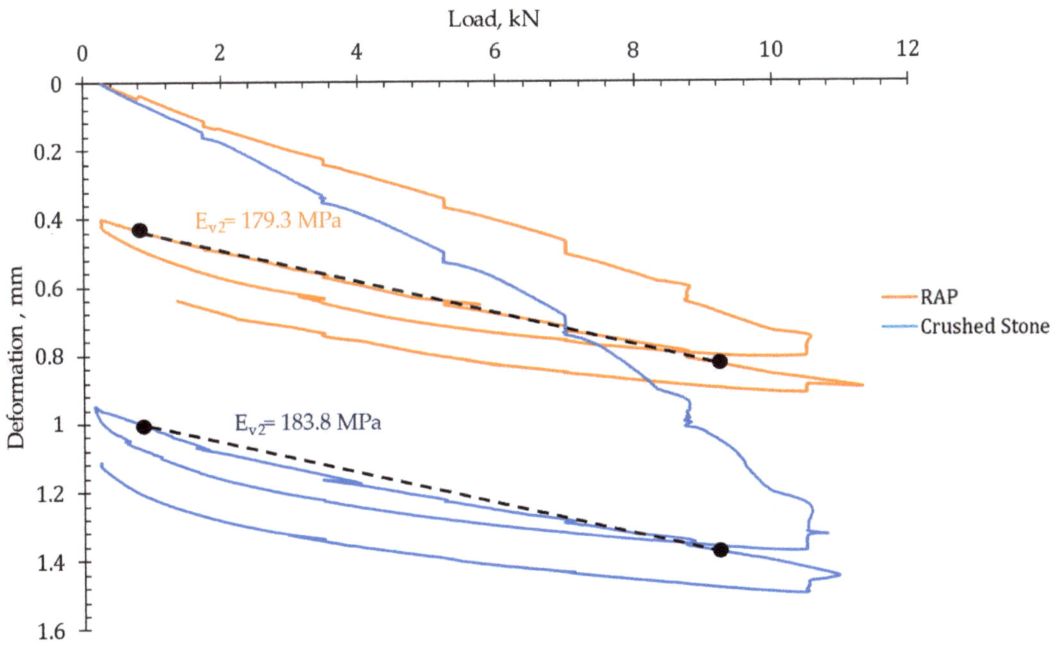

Figure 8. Static plate load test results.

In summary, the static load plate test suggests that, despite being a layer composed of smaller-sized material with less continuity in its particle size distribution, the residual asphalt binder contributes to the resistance capacity. The static plate load results further underscore the feasibility of using RAP as a base layer material, particularly for applications where deformation and load-bearing capacity are critical.

Similarly, Figure 9 presents the results of load-bearing capacity under dynamic loading (last cycle). In this case, the total deformation after 10 loading cycles is very similar for both materials. This finding is consistent with other studies reporting the suitability of RAP for dynamic applications when compacted at optimal moisture levels [62,63].

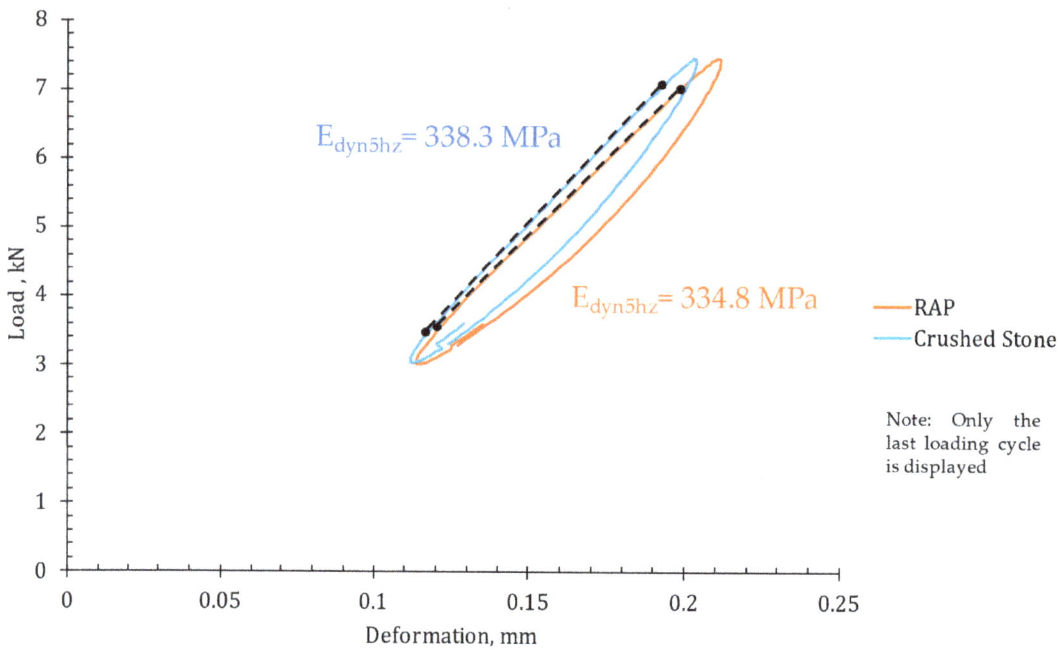

Figure 9. Dynamic plate load test results.

In terms of bearing capacity, the vertical elastic modulus (Edyn5hz) reached by the recycled materials is comparable to that of the material commonly used for these applications under this load condition, with minimal differences observed. The load-bearing capacity of the base layer is essential for effectively distributing stress to the underlying sublayers and minimizing settlement.

3.3. Permanent Deformation

The results of the wheel-tracking test, simulating the traveling of tire wheels over the layers of each material is presented in Figure 10, in this case, represents the evolution of layer deformation under the continuous passage of the wheel, presenting values of wheel-tracking slope (WTS, measured in millimeters of rut depth per 1000 wheel passes), mean proportional rut depth (PRD in %), and the depth of the rut (RD in mm). The results indicate that in this case, the recycled materials exhibit better performance than the reference material, suggesting that these recycled materials may experience lower total deformations during their application, Specifically, reductions of over 95% in the RD value were observed compared to the reference material.

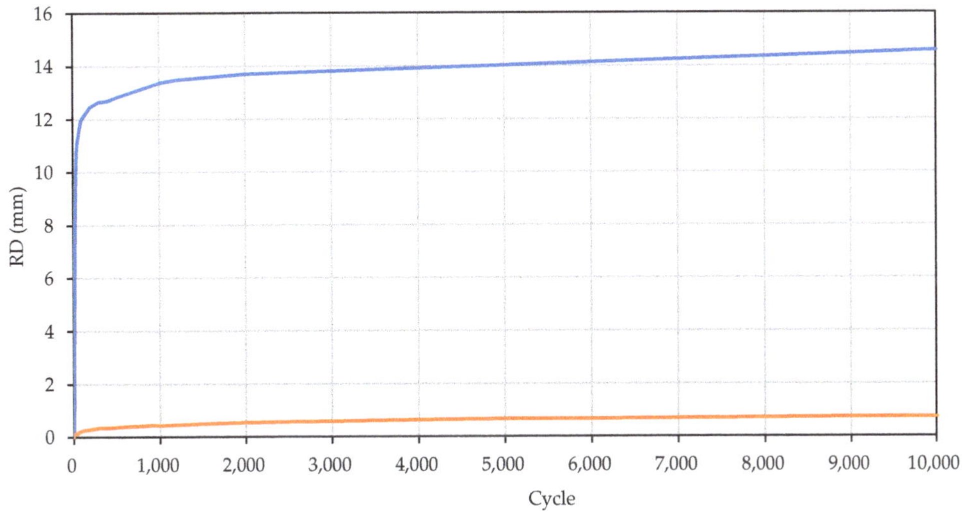

	Crushed Stone	RAP
WTS AIR (mm/1000 Cycles)	0.11	0.02
PRD (%)	18.21	0.90
RD (mm)	14.56	0.72

Figure 10. Wheel tracking test results.

However, some authors have reported higher rutting in RAP-based materials [68]. This discrepancy in findings may be attributed to the inherent heterogeneity of RAP, coupled with its physical properties, which play a critical role in its final performance. Factors such as bitumen content, gradation, and other physical characteristics significantly influence the material's behavior under load application. The values obtained in this study, while higher than those reported by other authors [65], align with findings from other research that highlight the consistent behavior of RAP under controlled conditions [69,70]. Furthermore, the recycled material demonstrated stable performance throughout the test, contrasting with the reference material, which exhibited rapid settlement during the initial cycles before reaching stabilization.

3.4. Permeability

The constant head permeability test was conducted until the water flow through the material stabilized. In this case, the values of vertical permeability coefficients (K) were calculated for each of the studied materials, obtaining values of 9.76×10^{-6} m/s for the recycled material and 8.24×10^{-6} m/s for the reference material; however, values were considerably higher than those reported in other studies [71]. It is important to note that in those studies, the authors incorporated fillers, such as cement, at approximately 3% in unbound base layers, aiming to reduce voids and improve permeability coefficients.

Overall, it is observed that the recycled material has a slightly lower capacity to protect the lower layers of the substructure. However, all materials exhibited permeability coefficients suitable for their application. This parameter is closely related to gradation, so having continuous gradation contributes significantly to the material's ability to protect the structure [72,73].

3.5. *Full-Scale Laboratory Test Box*

Based on the previous results, a full-scale test box was conducted to simulate operational and confinement conditions for the RAP and reference base layer. Tests were carried out on these materials to assess their load-bearing capacity and settlement under real conditions expected during the application of this solution.

3.5.1. Bearing Capacity of Full-Scale Texting Box

Figure 11 shows the results of elastic modulus under (a) static load and (b) dynamic load for the materials under study. These indicate that the load-bearing capacity of the recycled materials was slightly higher than that of the reference material, both under dynamic and static loads. In Figure 11a, the values of the elastic modules for the first and second cycles in the static plate load test are presented, along with the relationship between them. Overall, the modulus of elasticity in the second cycle of load was about 8.3% higher in the recycled material than in the reference material. However, it can be observed that the relationship between the modulus of elasticity in the second cycle and that of the first cycle was better in the case of the reference material, although both were close to 1. The total deformation after both cycles is similar for both materials. The observed load-bearing capacity value is slightly lower than that reported in the previous section; however, it tends to be more realistic due to the closer approximation to in-service confinement conditions.

Moreover, the elasticity modulus values under dynamic loads are higher for the recycled material compared to the reference material, confirming better load-bearing capacity under these conditions. Furthermore, the total deformation observed in the reference material is greater than that of the recycled material, indicating a higher settlement tendency.

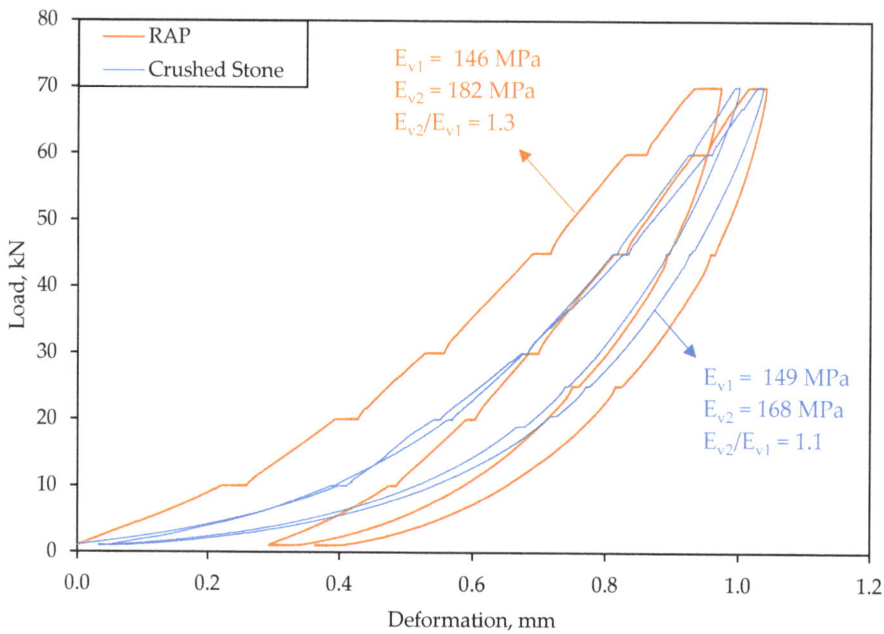

(**a**) Static loading plate

Figure 11. *Cont.*

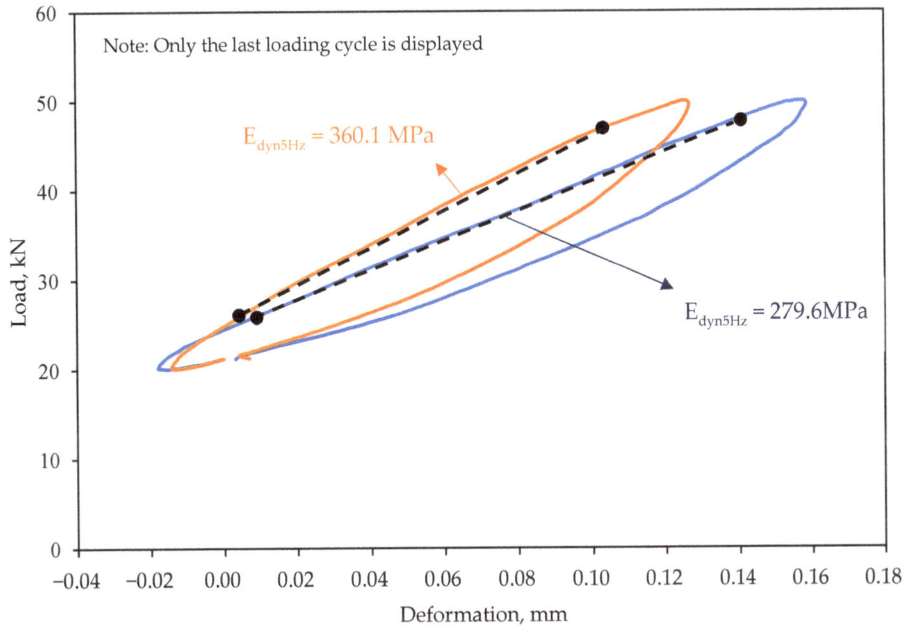

(**b**) Dynamic loading plate

Figure 11. Bearing capacity.

It is important to highlight that these findings are consistent with the results obtained at the specimen scale in previous sections, demonstrating the reproducibility of laboratory results when applied to full-scale test scenarios. This consistency further validates the suitability of RAP-based granular layers in infrastructure applications. Similar conclusions were drawn from previous full-scale test box studies, where the mechanical performance of RAP-based materials was assessed under real operational confinement conditions [74,75].

3.5.2. Influence on the Behavior of the Track Section

Figure 12 illustrates the evolution of permanent deformation in each of the studied materials after 110,000 loading cycles. Under more realistic confinement conditions, the recycled material exhibited lower permanent deformations than those observed in the reference material, in this case, with a reduction of 22% in total permanent deformation. The results indicate that the deformation slope of the recycled material in the initial stage (first 500 cycles) was higher, but once consolidated, the deformation slope decreased, resulting in a smaller final deformation.

The observed reduction in total settlement highlights the importance of stiffer base layers, as an increase in substructure stiffness leads to decreased settlements and enhanced resistance over time for the entire track [76]. This translates into lower maintenance requirements and a reduction in infrastructure management costs. The improved performance of the recycled material under cyclic loading suggests that RAP-based layers can enhance long-term durability, minimizing permanent deformation and potentially extending the service life of the pavement with reduced maintenance interventions.

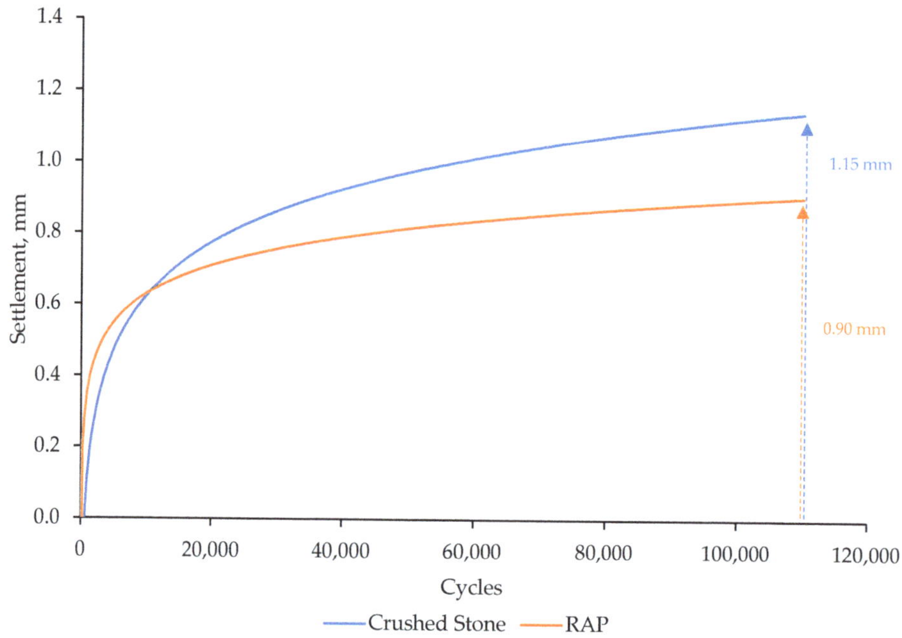

Figure 12. Permanent deformation.

4. Conclusions

The main objective of this study is to analyze the feasibility of using reclaimed asphalt pavement (RAP) material as a granular base in road infrastructure projects using advanced laboratory tests and full-scale laboratory test boxes to simulate real conditions. Based on the results obtained from laboratory tests characterizing different RAP types and evaluating the influence of various design parameters, the following conclusions can be drawn:

- The recycled materials analyzed in this study exhibited appropriate mechanical characteristics to be used as aggregates in base layers for road structures.
- The results demonstrate that the use of RAP enhances the internal cohesion and mechanical resistance of granular materials used in road-based applications. This improvement is attributed to the bonding capacity provided by the residual binder, which becomes activated at high ambient temperatures. This approach resulted in a 65% increase in compressive strength and up to a 126% increase in indirect tensile strength compared to reference materials.
- The load-bearing capacity of RAP-based materials was superior to that of conventional granular bases under the proposed compaction conditions, achieving up to 30% higher modulus of elasticity below static and dynamic loads.
- Despite its more discontinuous gradation, RAP demonstrated adequate permeability characteristics, ensuring sufficient protection for the lower layers of the pavement structure against water infiltration.
- In permanent deformation tests, RAP-based materials exhibited superior performance, with up to a 22% reduction in total permanent deformation compared to natural aggregates, which translates into lower maintenance needs over the pavement's service life.

Based on these conclusions, it has been demonstrated that RAP can be effectively used in high proportions as an unbound base layer, provided that appropriate compaction techniques are employed to facilitate the activation of the residual asphalt binder. Nonetheless, it is important to remark that although this study includes different RAP sources, it is difficult to establish universally applicable conclusions due to the heterogeneity of recycled RAP materials, as variations in aggregate type, binder properties, and asphalt content significantly influence their mechanical behavior.

Thus, further research is needed to assess the environmental and economic viability of RAP in granular-based applications. This includes comprehensive studies on life cycle assessments and cost analyses that consider potential reductions or increases in layer thickness to optimize performance and sustainability. A comparative evaluation of RAP reuse in different pavement layers, including asphalt mixtures and granular bases, is essential to determine the most sustainable and efficient application. Future studies should also focus on evaluating long-term field performance to validate laboratory findings and ensure practical implementation.

Author Contributions: Conceptualization, M.S.-S. and M.C.R.-G.; methodology, M.S.-S.; software, O.G.-B. and A.G.; validation, A.G., formal analysis, O.G.-B. and A.G.; investigation, O.G.-B. and A.G.; resources, M.S.-S.; data curation, O.G.-B. and A.G.; writing—original draft preparation, O.G.-B.; writing—review and editing, M.S.-S.; visualization, F.M.-N.; supervision, M.S.-S. and M.C.R.-G.; project administration, M.S.-S.; funding acquisition, M.S.-S. and F.M.-N. All authors have read and agreed to the published version of the manuscript.

Funding: The present study has been conducted within the framework of two research projects: one project in collaboration with the company GUAMAR S.A., entitled "Viability of using RAP as granular base in linear constructions"; and another project with acronym "ECO-WAVE" and titled "Development of Sustainable and anti-vibration sub-ballast for railway" (PID2022-142915OB-IOO), funded by the Ministry of Science, Innovation and University of Spain (MICIU/AEI/10.13039/501100011033) and by the European Union (FEDER).

Institutional Review Board Statement: Not Applicable.

Informed Consent Statement: Not Applicable.

Data Availability Statement: The data that supports the findings of this study are available from the author upon reasonable request.

Conflicts of Interest: The authors declare that there is no conflict of interest.

References

1. Elliot, T.; Carter, A.; Ghattuwar, S.; Levasseur, A. Environmental Impacts of Road Pavement Rehabilitation. *Transp. Res. D Transp. Environ.* **2023**, *118*, 103720. [CrossRef]
2. Garraín, D.; Lechón, Y. Environmental Footprint of a Road Pavement Rehabilitation Service in Spain. *J. Environ. Manag.* **2019**, *252*, 109646. [CrossRef] [PubMed]
3. Pradhan, S.K.; Biswal, G. Utilization of Reclaimed Asphalt Pavement (RAP) as Granular Sub-Base Material in Road Construction. *Mater. Today Proc.* **2022**, *60*, 288–293. [CrossRef]
4. Yaro, N.S.A.; Sutanto, M.H.; Baloo, L.; Habib, N.Z.; Usman, A.; Yousafzai, A.K.; Ahmad, A.; Birniwa, A.H.; Jagaba, A.H.; Noor, A. A Comprehensive Overview of the Utilization of Recycled Waste Materials and Technologies in Asphalt Pavements: Towards Environmental and Sustainable Low-Carbon Roads. *Processes* **2023**, *11*, 2095. [CrossRef]
5. Nemade, P.A.; Pasla, D.; Chandrappa, A.K. Durability Assessment of Concrete with Natural and Linz Donawitz Slag as Coarse Aggregates. *Constr. Build. Mater.* **2023**, *400*, 132617. [CrossRef]
6. Mohammed, S.I.; Najim, K.B. Mechanical Strength, Flexural Behavior and Fracture Energy of Recycled Concrete Aggregate Self-Compacting Concrete. *Structures* **2020**, *23*, 34–43. [CrossRef]
7. Sol-Sánchez, M.; Moreno-Navarro, F.; Rubio-Gámez, M.C. Viability Analysis of Deconstructed Tires as Material for Rail Pads in High-Speed Railways. *Mater. Des.* **2014**, *64*, 407–414. [CrossRef]
8. Marrero, M.; Rivero-Camacho, C.; Martínez-Rocamora, A.; Alba-Rodríguez, D.; Lucas-Ruiz, V. Holistic Assessment of the Economic, Environmental, and Social Impact of Building Construction. Application to Housing Construction in Andalusia. *J. Clean Prod.* **2024**, *434*, 140170. [CrossRef]

9. European Commission. *LEVEL (S) Taking Action on the Total Impact of the Construction Sector*; European Union (EU): Brussels, Belgium, 2019.
10. Guerrero-Bustamante, O.; Camargo, R.; Dawd, I.; Duque, J.; Polo-Mendoza, R.; Gálvis, J.; Díaz, J.; Daza, O.; Cucunuba, J.; Acosta, C. Implementation of Crumb Rubber (CR) in Road Pavements: A Comprehensive Literature Review. *Infrastructure* **2024**, *9*, 223. [CrossRef]
11. Arulrajah, A.; Jegatheesan, P.; Disfani, M.; Bo, M. Geotechnical and Geo-Environmental Properties of Recycled Construction and Demolition Materials in Pavement Subbase Applications. *J. Mater. Civ. Eng.* **2013**, *25*, 1077–1088. [CrossRef]
12. Liang, Y.; Ye, Z.; Vernerey, F.; Xi, Y. Development of Processing Methods to Improve Strength of Concrete with 100% Recycled Coarse Aggregate. *J. Mater. Civ. Eng.* **2013**, *27*, 4014163. [CrossRef]
13. Ghanbari, M.; Abbasi, A.M.; Ravanshadnia, M. Production of Natural and Recycled Aggregates: The Environmental Impacts of Energy Consumption and CO2 Emissions. *J. Mater. Cycles Waste Manag.* **2018**, *20*, 810–822. [CrossRef]
14. Wang, T.; Zhao, X.; Zheng, L.; Mao, C.; Wang, L.; Falchetto, A.C.; Guo, D. Hot In-Place Recycled Asphalt Mixtures: RAP Analysis, Compaction Characteristics and Field Evaluation. *Sustainability* **2024**, *16*, 1064. [CrossRef]
15. Yu, J.; Lin, Z.; Zou, G.; Yu, H.; Leng, Z.; Zhang, Y. Long-Term Performance of Recycled Asphalt Mixtures Containing High RAP and RAS. *J. Road Eng.* **2024**, *4*, 36–53. [CrossRef]
16. Moreno-Navarro, F.; Sierra-Carrillo de Albornoz, F.J.; Sol-Sánchez, M.; Rubio-Gámez, M.C. MASAI: Sustainable, Automated and Intelligent Asphalt Materials. The Way to the next Generation of Asphalt Pavements. *Road Mater. Pavement Des.* **2023**, *24*, 486–505. [CrossRef]
17. Zaumanis, M.; Arraigada, M.; Wyss, S.A.; Zeyer, K.; Cavalli, M.C.; Poulikakos, L.D. Performance-Based Design of 100% Recycled Hot-Mix Asphalt and Validation Using Traffic Load Simulator. *J. Clean. Prod.* **2019**, *237*, 117679. [CrossRef]
18. Hoppe, E.J.; Lane, D.S.; Fitch, G.M.; Shetty, S. *Feasibility of Reclaimed Asphalt Pavement (RAP) Use as Road Base and Subbase Material*; Virginia Center for Transportation Innovation and Research: Charlottesville, VA, USA, 2015.
19. Guerrero-Bustamante, O.; Guillen, A.; Moreno-Navarro, F.; Rubio-Gámez, M.C.; Sol-Sánchez, M. Toward Sustainable Railway Track Foundation: Recycling RAP from Road Asphalt Pavements. *Constr. Build. Mater.* **2024**, *439*, 137436. [CrossRef]
20. Thakur, J.K.; Han, J. Recent Development of Recycled Asphalt Pavement (RAP) Bases Treated for Roadway Applications. *Transp. Infrastruct. Geotechnol.* **2015**, *2*, 68–86. [CrossRef]
21. Luo, C. High Performance Granular Base and Subbase Materials Incorporating Reclaimed Asphalt Concrete Pavement. Doctoral Thesis, McMaster University, Hamilton, ON, Canada, 2014.
22. Arshad, M.; Ahmed, M.F. Potential Use of Reclaimed Asphalt Pavement and Recycled Concrete Aggregate in Base/Subbase Layers of Flexible Pavements. *Constr. Build. Mater.* **2017**, *151*, 83–97. [CrossRef]
23. Coban, H.S.; Cetin, B.; Ceylan, H.; Edil, T.B.; Likos, W.J. Evaluation of Long-Term Performance of Recycled Aggregate Base (RAB) Layers and Optimization of Their Design Thicknesses. *Road Mater. Pavement Des.* **2023**, *24*, 1310–1329. [CrossRef]
24. Yuan, D.; Nazarian, S.; Hoyos, L.; Puppala, A. Evaluation and Mix Design of Cement-Treated Base Materials with High Content of Reclaimed Asphalt Pavement. *Transp. Res. Rec. J. Transp. Res. Board* **2011**, *2212*, 110–119. [CrossRef]
25. Polo-Mendoza, R.; Mora, O.; Duque, J.; Turbay, E.; Martinez-Arguelles, G.; Fuentes, L.; Guerrero, O.; Perez, S. Environmental and Economic Feasibility of Implementing Perpetual Pavements (PPs) against Conventional Pavements: A Case Study of Barranquilla City, Colombia. *Case Stud. Constr. Mater.* **2023**, *18*, e02112. [CrossRef]
26. Attia, M.; Abdelrahman, M. Effect of State of Stress on the Resilient Modulus of Base Layer Containing Reclaimed Asphalt Pavement. *Road Mater. Pavement Des.* **2011**, *12*, 79–97. [CrossRef]
27. Rodríguez-Fernández, I.; Lastra-González, P.; Indacoechea-Vega, I.; Castro-Fresno, D. Technical Feasibility for the Replacement of High Rates of Natural Aggregates in Asphalt Mixtures. *Int. J. Pavement Eng.* **2021**, *22*, 940–949. [CrossRef]
28. Coelho, L.M.; Guimarães, A.C.R.; de Azevedo, A.R.G.; Monteiro, S.N. Sustainable Reclaimed Asphalt Emulsified Granular Mixture for Pavement Base Stabilization: Prediction of Mechanical Behavior Based on Repeated Load Triaxial Tests. *Sustainability* **2024**, *16*, 5335. [CrossRef]
29. Bustos, G. *Pliego de Prescripciones Técnicas Generales Para Obras de Carreteras y Puentes PG-3*, 4th ed.; Ediciones Liteam: Madrid, Spain, 2004.
30. *UNE EN 933-3:2023*; Part 3: Determination of Particle Shape—Flakiness Index. AENOR—Asociación Española de Normalización y Certificación NORMA: Madrid, Spain, 2012.
31. *UNE EN 933-5:2023*; Part 5: Determination of Percentage of Crushed particles in Coarse and All-in Natural Aggregates. AENOR—Asociación Española de Normalización y Certificación NORMA: Madrid, Spain, 2023.
32. *UNE-EN 933-8:2012+A1*; Part 8: Assess-ment of Fines—Sand Equivalent Test. AENOR—Asociación Española de Normalización y Certificación NORMA: Madrid, Spain, 2015.
33. *UNE EN 1097-2:2021*; Part 2: Methods for the Determination of Resistance to Fragmentation. AENOR—Asociación Española de Normalización y Certificación NORMA: Madrid, Spain, 2021.

34. *UNE-EN 1097-6*; Part 6: Determination of Particle Density and Water Absorption. AENOR—Asociación Española de Normalización y Certificación NORMA: Madrid, Spain, 2022.
35. *UNE EN 933-1:2012*; Part 1: Determination of Particle Size Distribution—Sieving Method. AENOR—Asociación Española de Normalización y Certificación NORMA: Madrid, Spain, 2012.
36. *UNE 103501:1994*; Geotecnia. Ensayo de Compactación. Proctor Modificado. AENOR—Asociación Española de Normalización y Certificación NORMA: Madrid, Spain, 1994.
37. *UNE-EN 12697-1:2013*; Bituminous Mixtures. Test Methods for Hot Mix Asphalt. Part 1: Soluble Binder Content. AENOR—Asociación Española de Normalización y Certificación NORMA: Madrid, Spain, 2013.
38. *UNE-EN 1426:2015*; Bitumen and Bituminous Binders. Determination of Needle Penetration. AENOR—Asociación Española de Normalización y Certificación NORMA: Madrid, Spain, 2015.
39. *UNE-EN 1427:2015*; Bitumen and Bituminous Binders. Determination of the Softening Point. Ring and Ball Method. AENOR—Asociación Española de Normalización y Certificación NORMA: Madrid, Spain, 2015.
40. Pirozzolo, L.; Sol-Sánchez, M.; Moreno-Navarro, F.; Martínez-Montes, G.; Rubio-Gámez, M.C. Evaluation of Bituminous Sub-Ballast Manufactured at Low Temperatures as an Alternative for the Construction of More Sustainable Railway Structures. *Mater. De Construcción* **2017**, *67*, e128. [CrossRef]
41. Taziani, E.A.; Toraldo, E.; Mariani, E.; Crispino, M. Investigation on the Compaction Temperature Effect on Reactivating Aged Bitumen of 100% RAP Mixtures. In *Pavement and Asset Management*; Taylor & Francis Group: Abingdon, UK, 2019; ISBN 9780429264702.
42. Gong, X.; Liu, Q.; Lv, Y.; Chen, S.; Wu, S.; Ying, H. A Systematic Review on the Strategies of Reducing Asphalt Pavement Temperature. *Case Stud. Constr. Mater.* **2023**, *18*, e01852. [CrossRef]
43. Qin, Y.; Zhang, X.; Tan, K.; Wang, J. A Review on the Influencing Factors of Pavement Surface Temperature. *Environ. Sci. Pollut. Res.* **2022**, *29*, 67659–67674. [CrossRef]
44. Coelho, L.M.; Guimarães, A.C.R. Effect of Compaction Temperature on the Mechanical Behavior of Granular Base Containing RAP. In Proceedings of the 5th International Conference on Transportation Geotechnics (ICTG), Sydney, Australia, 20–22 November 2024; Volume 5, pp. 129–137.
45. Costa, J.O.; Borges, P.H.R.; dos Santos, F.A.; Bezerra, A.C.S.; Van den bergh, W.; Blom, J. Cementitious Binders and Reclaimed Asphalt Aggregates for Sustainable Pavement Base Layers: Potential, Challenges and Research Needs. *Constr. Build. Mater.* **2020**, *265*, 120325. [CrossRef]
46. Tauste, R.; Moreno-Navarro, F.; Sol-Sánchez, M.; Rubio-Gámez, M.C. The Effect of the Nature of Rejuvenators on the Rheological Properties of Aged Asphalt Binders. In Proceedings of the RILEM 252-CMB Symposium: Chemo-Mechanical Characterization of Bituminous Materials, Braunschweig, Germany, 17–18 September 2018; Springer: Cham, Switzerland, 2019; pp. 220–225.
47. Moreno-Navarro, F.; Sol-Sánchez, M.; García-Travé, G.; Tauste-Martínez, R.; Rubio-Gámez, M.C.; Rodriguez-López, C.; Hernández-Pérez, M.; Gambín-Peñalver, J. Laboratory Study of Soil-Cement Manufactured with RAP for Its Application as a Sustainable and High-Performance Base in Asphalt Pavements. In *Eleventh International Conference on the Bearing Capacity of Roads, Railways and Airfields*; CRC Press: London, UK, 2022; Volume 3, pp. 313–321.
48. Navarro, F.M.; Gámez, M.C.R. Influence of Crumb Rubber on the Indirect Tensile Strength and Stiffness Modulus of Hot Bituminous Mixes. *J. Mater. Civ. Eng.* **2012**, *24*, 715–724. [CrossRef]
49. *NORMA UNE 103400:1993*; Simple Compression Test on Soil Specimens. AENOR-Asociación Española de Normalización y Certificación: Madrid, Spain, 1993.
50. *NORMA UNE-EN 12697-23:2018*; Part 23: Determination of the Indirect Tensile Strength of Bituminous Specimens. AENOR-Asociación Española de Normalización y Certificación: Madrid, Spain, 2018.
51. *NORMA UNE 103808:2006*; Ensayo de Carga Vertical de Suelos Mediante Placa Estática. AENOR-Asociación Española de Normalización y Certificación: Madrid, Spain, 2006.
52. *NORMA UNE 103807-2:2021*; Ensayo de Carga Vertical de Suelos Mediante Placa Dinámica. AENOR-Asociación Española de Normalización y Certificación: Madrid, Spain, 2021.
53. *UNE EN 12697-22:2022*; Bituminous Mixtures—Test Methods—Part 22: Wheel Tracking. AENOR-Asociación Española de Normalización y Certificación: Madrid, Spain, 2022.
54. *NORMA UNE 103403:1999*; Determinación de La Permeabilidad de Una Muestra de Suelo—Método de Carga Constante. AENOR-Asociación Española de Normalización y Certificación: Madrid, Spain, 1999.
55. *UNE-EN 13108-2:2007*; Bituminous Mixtures—Material Specifications—Part 2: Asphalt Concrete for Very Thin Layers. AENOR-Asociación Española de Normalización y Certificación: Madrid, Spain, 2007.
56. Sol-Sánchez, M.; Moreno-Navarro, F.; Carmen Rubio-Gámez, M. Large-Scale Testing Box: A Tool to Evaluate the Structural Performance of Railway Tracks. In *Railroad Ballast Testing and Properties*; ASTM International100 Barr Harbor Drive: West Conshohocken, PA, USA, 2018; pp. 191–207.

57. Alam, T.B.; Abdelrahman, M.; Schram, S.A. Laboratory Characterisation of Recycled Asphalt Pavement as a Base Layer. *Int. J. Pavement Eng.* **2010**, *11*, 123–131. [CrossRef]
58. Tauste, R.; Hidalgo, A.E.; García, G.M.; Moreno-Navarro, F.; Rubio-Gámez, M.C. Understanding the Influence of Filler Type and Asphalt Binder Content on the Moisture and Fatigue Resistance of Asphalt Mortars. *Mater. De Construcción* **2022**, *72*, e301. [CrossRef]
59. Moreno-Navarro, F.; Sol-Sánchez, M.; Tomás-Fortún, E.; Rubio-Gámez, M.C. High-Modulus Asphalt Mixtures Modified with Acrylic Fibers for Their Use in Pavements under Severe Climate Conditions. *J. Cold Reg. Eng.* **2016**, *30*, 04016003. [CrossRef]
60. Brand, A.S.; Roesler, J.R.; Al-Qadi, I.L.; Shangguan, P. *Fractionated Reclaimed Asphalt Pavement (FRAP) as a Coarse Aggregate Replacement in a Ternary Blended Concrete Pavement*; CORE: Downers Grove, IL, USA, 2012.
61. Su, Y.-M.; Hossiney, N.; Tia, M.; Bergin, M. Mechanical Properties Assessment of Concrete Containing Reclaimed Asphalt Pavement Using the Superpave Indirect Tensile Strength Test. *J. Test Eval.* **2014**, *42*, 912–920. [CrossRef]
62. Arshad, M. Development of a Correlation between the Resilient Modulus and CBR Value for Granular Blends Containing Natural Aggregates and RAP/RCA Materials. *Adv. Mater. Sci. Eng.* **2019**, *2019*, 1–16. [CrossRef]
63. Dong, Q.; Huang, B. Laboratory Evaluation on Resilient Modulus and Rate Dependencies of RAP Used as Unbound Base Material. *J. Mater. Civ. Eng.* **2014**, *26*, 379–383. [CrossRef]
64. Kumar Pradhan, S.; Das, U.; Ranjan Patra, A. Utilization of Reclaimed Asphalt Pavement (RAP) Materials in HMA Mixtures for Flexible Pavement Construction. *Mater. Today Proc.* **2023**. [CrossRef]
65. Jacobs, G.; Couscheir, K.; Hernando, D.; Moins, B.; Vansteenkiste, S.; Tanghe, T.; Duerinckx, B.; Van den bergh, W. Mechanical Performance of Asphalt Base Layers with High RAP Content and Recycling Agents. *Road Mater. Pavement Des.* **2024**, *25*, 84–94. [CrossRef]
66. Hidalgo-Signes, C.; Martínez-Fernández, P.; Garzón-Roca, J.; Insa-Franco, R. Analysis of the Bearing Capacity of Unbound Granular Mixtures with Rubber Particles from Scrap Tyres When Used as Sub-Ballast. *Mater. De Construcción* **2016**, *66*, 105. [CrossRef]
67. Seferoğlu, A.G.; Seferoğlu, M.T.; Akpinar, M.V. Experimental Study on Cement-Treated and Untreated RAP Blended Bases: Cyclic Plate Loading Test. *Constr. Build. Mater.* **2018**, *182*, 580–587. [CrossRef]
68. Arshad, A.K.; Abd Karim, Z.; Shaffie, E.; Hashim, W.; Abd Rahman, Z. Marshall Properties and Rutting Resistance of Hot Mix Asphalt with Variable Reclaimed Asphalt Pavement (RAP) Content. *IOP Conf. Ser. Mater. Sci. Eng.* **2017**, *271*, 012078. [CrossRef]
69. Howard, I.L.; Doyle, J.D.; Cox, B.C. Merits of Reclaimed Asphalt Pavement-Dominated Warm Mixed Flexible Pavement Base Layers. *Road Mater. Pavement Des.* **2013**, *14*, 106–128. [CrossRef]
70. Zhang, J.; Sun, H.; Jiang, H.; Xu, X.; Liang, M.; Hou, Y.; Yao, Z. Experimental Assessment of Reclaimed Bitumen and RAP Asphalt Mixtures Incorporating a Developed Rejuvenator. *Constr. Build. Mater.* **2019**, *215*, 660–669. [CrossRef]
71. Seferoğlu, A.G.; Seferoğlu, M.T.; Akpınar, M.V. Investigation of the Effect of Recycled Asphalt Pavement Material on Permeability and Bearing Capacity in the Base Layer. *Adv. Civ. Eng.* **2018**, *2018*, 1–6. [CrossRef]
72. Koohmishi, M.; Azarhoosh, A. Assessment of Permeability of Granular Drainage Layer Considering Particle Size and Air Void Distribution. *Constr. Build. Mater.* **2021**, *270*, 121373. [CrossRef]
73. Kandlavath, H.N.; Chowdhury, P.S.; Reddy, M.A. Evaluation of Horizontal Permeability Characteristics of Granular Subbase Material. *Transp. Res. Procedia* **2020**, *48*, 3725–3733. [CrossRef]
74. Sol-Sánchez, M.; Moreno-Navarro, F.; Rubio-Gámez, M.C.; Manzo, N.; Fontseré, V. Full-Scale Study of Neoballast Section for Its Application in Railway Tracks: Optimization of Track Design. *Mater. Struct.* **2018**, *51*, 43. [CrossRef]
75. d'Angelo, G.; Sol-Sánchez, M.; Moreno-Navarro, F.; Lo Presti, D.; Thom, N. Use of Bitumen-Stabilised Ballast for Improving Railway Trackbed Conventional Maintenance. *Géotechnique* **2018**, *68*, 518–527. [CrossRef]
76. Sol-Sánchez, M.; Pirozzolo, L.; Moreno-Navarro, F.; Rubio-Gámez, M.C. A Study into the Mechanical Performance of Different Configurations for the Railway Track Section: A Laboratory Approach. *Eng. Struct.* **2016**, *119*, 13–23. [CrossRef]

materials MDPI

Article

Effects of Complex Environmental Conditions on Fatigue Self-Healing Properties of Fast-Melting SBS-Modified Asphalt

Jinchao Yue [1], Jiahao Fu [1], Yihan Wang [1], Yang Ming [2], Guoqi Tang [3] and Ruixia Li [1,*]

1 School of Water Conservancy and Transportation, Zhengzhou University, Zhengzhou 450001, China; yuejc@zzu.edu.cn (J.Y.); fujh38542426@163.com (J.F.); 17513162919@163.com (Y.W.)
2 Henan Transport Investment Group Co., Ltd., Zhengzhou 450046, China; 13087025558@163.com
3 Guolu Gaoke Engineering Technology Institute Co., Ltd., Beijing 100083, China; tang12341102@163.com
* Correspondence: ruixiali031@163.com

Abstract: Asphalt pavements are prone to various distresses under complex environmental influences during service, which significantly affects their fatigue life. This study conducted complex environmental simulation tests, including pressure aging, ultraviolet (UV) aging, and coupling effects with salt solutions at different concentrations. A dynamic shear rheometer (DSR) was employed to perform frequency sweep tests, linear amplitude sweep (LAS) tests, and fatigue–healing–fatigue tests. The fatigue self-healing properties of fast-melting SBS (SBS-T)-modified asphalt were evaluated based on the viscoelastic continuous damage theory. The results indicate that coupled aging effects significantly increase the viscoelastic characteristic parameters of SBS-T-modified asphalt, with more elastic components transforming into viscous components. Compared to other aging effects, the coupled pressure-UV-salt solution condition induces the most severe degradation in the fatigue durability of SBS-T-modified asphalt. Simultaneously, the self-healing capability of aged asphalt is also reduced. Specifically, with increasing strain, more complex aging conditions lead to the faster deterioration of asphalt fatigue life and lower self-healing capacity. While asphalt demonstrates measurable fatigue life restoration through self-healing, the synergistic coupling of salt solution exposure and multi-factor aging significantly compromises both the absolute fatigue resistance and the relative recovery efficiency.

Keywords: fast-melting SBS; complex environmental; fatigue self-healing properties; modified asphalt

1. Introduction

Asphalt pavements are persistently subjected to vehicular loading in conjunction with environmental stressors, notably UV radiation, thermal oxidation, and moisture infiltration, during prolonged service periods, resulting in diverse distress manifestations such as fatigue cracking and thermal-induced fracturing. Asphalt aging is recognized as the primary cause of the reduced durability and shortened service life of high-grade highways. The prevention or mitigation of crack propagation to extend the service life of asphalt pavements has emerged as a critical research frontier in pavement sustainability engineering [1]. Among various pavement distresses, fatigue damage is particularly crucial to the service performance of asphalt pavements, as fatigue life directly determines the pavement's service duration. Due to the unique characteristics of asphalt materials, their fatigue failure behavior is influenced by nonlinear viscoelastic effects, thixotropy, spatial hardening, and self-healing behavior [2]. Consequently, advancing the precision of asphalt fatigue life prediction necessitates the systematic incorporation of self-healing mechanisms into fatigue performance characterization frameworks.

The current research on the effects of external environments on the fatigue self-healing properties of asphalt materials primarily focuses on individual factors. However, in actual road service conditions, pavements are subjected to multiple simultaneous factors. Pressure, UV radiation, and saline environments can accelerate asphalt aging, significantly affecting its durability. Laboratory aging simulation tests, enabling precise parameter control and accelerated aging regimes, serve as standardized protocols to investigate asphalt aging mechanisms [3]. The Rolling Thin Film Oven Test (RTFOT) and Pressure Aging Vessel (PAV) are standardized protocols employed to mimic short-term oxidative aging and long-term thermomechanical degradation of asphalt binders, respectively. However, these accelerated aging protocols fall short of accurately reproducing in-service pavement environments owing to the exclusion of UV radiation and moisture-induced degradation mechanisms. Photooxidation is a major aging factor in regions with intense sunlight, particularly UV radiation from the solar spectrum, which significantly accelerates asphalt pavement aging [4]. Wang et al. established UV radiation as a dominant degradation mechanism in asphalt aging, inducing significant deterioration of fatigue resistance, elevation in creep stiffness, and reduction in failure strain [5]. Photooxidative pathways, when coupled with thermal oxidation kinetics, enable a more accurate replication of service-induced aging processes compared to conventional thermal aging protocols, positioning UV exposure as a critical parameter in comprehensive durability assessments [6].

In China's inland and coastal saline–alkali regions, harsh natural conditions make asphalt pavements and bridge decks more susceptible to various early distresses. When exposed to rainwater, salt components in the soil can adversely affect asphalt pavements in solution form [7]. Moreover, the water contacting asphalt pavements is not pure, as salts used for snow and ice melting can form damaging solutions. Coastal regions, where roads and sidewalks are predominantly asphalt-paved, face additional challenges from atmospheric salt, rainwater, and sea fog containing substantial sea salt, which degrade the pavement performance and service life [8–10]. However, the current asphalt pavement material design in China still relies on conventional empirical methods, rarely considering the combined effects of UV radiation, pressure, and salt solutions on durability, making it inadequate for specific environmental demands. There is an urgent need to investigate the deterioration dynamics of asphalt under synergistic environmental stressors, which is critical for advancing pavement sustainability.

Asphalt, particularly polymer-modified asphalt, is the primary material for expressway construction in China, and its viscoelastic response in complex environments fundamentally governs the structural integrity of asphalt concrete pavements and their long-term serviceability [11,12]. The SBS modifier has emerged as the primary choice for asphalt modification owing to its superior compatibility with base asphalt coupled with the ability to impart exceptional elasticity and enhanced high-temperature performance to modified asphalt systems [13]. However, traditional wet process SBS-modified asphalt production faces technical and management challenges, including segregation and aging issues that are difficult to monitor and control, significantly affecting performance [14,15]. Considering these limitations, the dry process, where modifiers are directly mixed with base asphalt in the mixer, shows superior advantages by eliminating the modification production process. This approach not only prevents aging caused by pre-mixing asphalt with modifiers during production but also reduces carbon emissions compared to conventional modification methods, making it more environmentally friendly [16]. The widespread adoption of dry modification technology is not only an industry necessity but also provides crucial technical support for green road construction. Research [17] on the pavement performance of SBS-T modifiers showed that all technical requirements met the specification standards, with some indicators surpassing those of wet process asphalt. With increasing applications of dry technology in practical engineering, it is essential to focus on SBS-T-modified asphalt performance to provide reference data for engineering applications.

In summary, to thoroughly investigate the combined effects of pressure, UV radiation, and salt solutions on the fatigue self-healing properties of SBS-T-modified asphalt, the current investigation conducted experiments under designed conditions of coupled pressure, UV, and saline environments. Using dynamic shear rheological tests and based on the viscoelastic continuous damage theory, the research predicts asphalt fatigue life to reveal the influence patterns of complex environments on asphalt's fatigue self-healing performance.

2. Materials and Methods

2.1. Materials

The base asphalt (70#:Penetration Grade 70, petroleum asphalt with a penetration range of 60–80 (0.1 mm) under standard test conditions at 25 °C) utilized in this investigation was procured from Beijing Guolu Gaoke Engineering Technology Institute Co., Ltd. The fundamental properties of the asphalt were tested in accordance with the "Standard Test Methods of Bitumen and Bituminous Mixtures for Highway Engineering". The detailed test results are presented in Table 1.

Table 1. Basic properties of the base asphalt [18].

Property	Value
Ductility (25 °C, cm)	>100
Penetration (25 °C, 0.1 mm)	72
Dynamic viscosity (135 °C, Pa·s)	0.68
Softening point (°C)	50.5

The SBS-T modifier used for preparing polymer-modified asphalt was also provided by Beijing Guolu Gaoke Engineering Technology Institute Co., Ltd. The technical specifications of the modifier are detailed in Table 2.

Table 2. Technical indices of the SBS-T.

Indexes	Results
Mass of a single particle (g)	0.25
Styrene percentage (%)	30~32
Ash (%)	0.25
Dispersion of dry mixing	No particle residue

2.2. Preparation of Fast-Melting SBS-Modified Asphalt

Previous studies have demonstrated that, for both radial and axial SBS-T-modified asphalt, when the modifier content exceeds 4%, it can meet all technical specifications required for construction-grade modified asphalt. Since the 4% dosage of radial SBS-T is commonly adopted in actual construction practices [19], this study utilized a 4% concentration of SBS-T modifier to prepare the modified asphalt, as illustrated in Figure 1. To align the properties of the laboratory-prepared asphalt with those of modified asphalt used in practical engineering, the following preparation protocol was adopted:

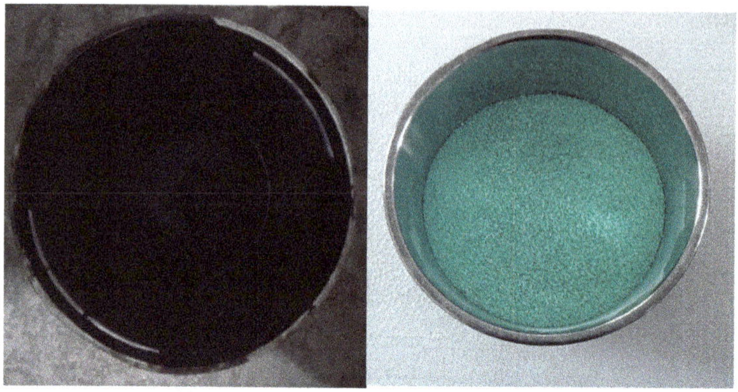

(**a**) (**b**)

Figure 1. (**a**) Fast-melting SBS-modified asphalt and (**b**) SBS-T.

(1) The base asphalt was subjected to thermal treatment at 180 °C under continuous mechanical agitation, followed by incremental incorporation of the predetermined dosage of SBS-T modifier while maintaining a rotational speed of 800 rpm to ensure homogeneous dispersion.

(2) The complete addition of SBS-T was accomplished within 2 min, following the principle of multiple small additions to ensure uniform dispersion and prevent agglomeration. Intermittent sampling for fluorescence microscopy analysis was conducted to verify phase compatibility and swelling completeness, thereby facilitating the complete swelling of SBS-T.

(3) Following complete incorporation of the SBS-T modifier, the system was subjected to high-shear homogenization at 5000 rpm for 40 min under strictly controlled thermal conditions (180–190 °C). The temperature was continuously monitored throughout the process to ensure proper modification, resulting in the final SBS-T-modified asphalt product.

2.3. Preparation of Salt Solutions

The salt compound selected for simulating saline environments in this study was sodium chloride (NaCl). Considering the actual salt composition in field conditions, salt solutions with concentrations of 6%, 8%, and 10% were formulated using NaCl crystals. These distinct concentration gradients were designed to simulate the effects of varying deicing salt dosages on the performance characteristics of SBS-T-modified asphalt.

2.4. Aging Process Simulation Tests

2.4.1. UV–Pressure Aging Test

To initiate the study, pour the SBS-T-modified asphalt sample evenly into a sample bottle with dimensions of 139.7 ±1.5 mm in height, 64 ±1.2 mm outer diameter, 2.4 ± 0.3 mm wall thickness, and 31.75 ±1.5 mm mouth diameter, containing 35 ± 0.5 g of asphalt. Place the sample bottle in a rotating thin-film oven maintained at a constant temperature of 163 ± 0.5 °C. The annular rack holding the sample bottles rotates at 15 ± 0.2 r/min while simultaneously injecting hot air at a flow rate of 4000 ± 200 mL/min. This process is maintained for 85 min. This procedure is designed to simulate the short-term aging process of asphalt. Following RTFOT conditioning to simulate short-term aging, the SBS-T-modified asphalt specimens were subjected to pressurized aging in a PAV for long term. Subsequently, the aged asphalt specimens were exposed to UV radiation in an ultraviolet aging chamber to replicate the synergistic interactions between thermal–oxidative degradation and UV-induced photoaging mechanisms.

From Figure 2, the ultraviolet aging system incorporated a 1.0 kW mercury–xenon arc lamp with spectral output spanning 345–355 nm. Samples were maintained at a 50 cm distance from the irradiation source, achieving a measured irradiance of 150 W/m^2 via the calibrated UV radiometer. Temperature control ensured a regulated internal chamber environment of 45 $\pm$ 1 °C. Based on previous studies [20,21], most researchers adopt UV aging durations of less than 15 days. Therefore, a 7-day UV exposure period was selected for this study, consistent with the literature. This duration ensures significant aging acceleration while balancing experimental efficiency and avoiding under-degradation risks [22].

Figure 2. UV aging chamber.

2.4.2. UV–Salt Solution Aging Test

The short-term aged asphalt was placed in an asphalt heating chamber and heated to a molten state (160 °C) to ensure sufficient fluidity for separation and weighing. Subsequently, 20 g of asphalt residue was weighed and poured into a tray to form a uniform thin film with a controlled thickness of $\leq$2 mm. These specimens were subsequently subjected to UV aging in the chamber using the procedure described in Section 2.4.1. To introduce saline environmental effects, the controlled daily spraying of prepared salt solutions at varying concentrations (6%, 8%, and 10%) was performed on the asphalt surface during UV exposure. Figure 3 illustrates the asphalt specimens after completing the coupled UV–salt solution aging process.

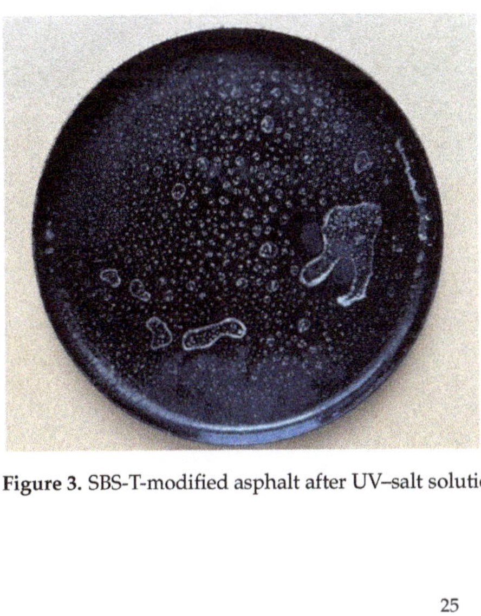

Figure 3. SBS-T-modified asphalt after UV–salt solution coupled aging.

2.4.3. UV–Pressure–Salt Solution Aging Test

A total of 50 g of short-term aged asphalt was placed in trays for long-term aging. Subsequently, 20 g of the long-term aged asphalt was weighed and placed in trays for UV aging. To avoid the influence of high temperatures on the salt solution, pre-configured salt solutions were sprayed daily onto the asphalt surface during the UV aging process. The pressure aging and UV aging procedures followed the same steps as described in the previous sections.

2.5. Experimental Protocols

2.5.1. Frequency Sweep Test

Frequency sweep rheometry serves as a fundamental technique for rheological characterization of asphalt binders' viscoelastic response. By applying varying loading frequencies and small strain levels at specific temperatures, the dynamic shear modulus of asphalt within its linear viscoelastic range can be determined.

A strain amplitude of 0.1% was applied within the linear viscoelastic regime of asphalt binders, which generally demonstrate linear behavior up to 15% strain. Rheological characterization was performed using a DSR under oscillatory loading frequencies spanning 0.2–30 Hz (12 discrete frequencies: 0.2, 0.4, 0.6, 0.8, 1, 2, 4, 6, 8, 10, 20, and 30 Hz) across three isothermal conditions: 15 °C, 25 °C, and 35 °C. Continuous monitoring captured the complex shear modulus (G^*) and phase angle (δ) evolution [23,24].

Implementing the time–temperature superposition principle (TTSP) with 25 °C as the reference temperature, horizontal shifting factors were computationally derived to superimpose complex modulus data acquired at 15 °C, 20 °C, 30 °C, and 35 °C, thereby constructing a unified master curve characterizing the asphalt's viscoelastic response across extended frequency–temperature domains. The shift factors were quantified under reference temperature conditions, and the reduced angular frequency used in the master curve is expressed as

$$log\ \omega_r = log\varphi_T + log\omega, \tag{1}$$

where ω is the angular frequency (rad/s), ω_r is the reduced angular frequency (rad/s), and φ_T is the shift factor.

2.5.2. Linear Amplitude Sweep (LAS) Test

The LAS test, based on the Viscoelastic Continuum Damage (VECD) theory [25], employs a strain-controlled loading mode. Nonlinear fitting of the stress–strain data obtained from the LAS test allows the application of the VECD model, enabling the derivation of the damage characteristic curve (DCC) [26]. This curve provides a foundation for predicting fatigue performance. The loading protocol is illustrated in Figure 4. Continuous monitoring of dynamic modulus evolution, phase angle variation, and nonlinear viscoelastic response enables rigorous fatigue life prediction through the VECD theory framework.

The VECD theory is a theoretical model used to describe the behavior of materials under dynamic loading. This model integrates viscoelasticity and continuum damage mechanics, enabling the characterization of strength and stiffness reduction under high strain rates, as well as the fracture and damage evolution in materials. This methodology demonstrates significant applicability in fatigue characterization of asphalt materials, particularly for evaluating the resistance to crack propagation under cyclic loading conditions. The DCC of a material can be determined through simple monotonic loading strength tests.

The DCC effectively describes the initiation and evolution of damage [27]. The damage evolution process is typically represented by Equations (2) and (3):

$$W^R = f\left(\varepsilon^R, S\right) = \frac{1}{2}\sigma\varepsilon^R, \quad (2)$$

$$\frac{dS}{dt} = \left(-\frac{\partial W^R}{\partial S}\right)^{\alpha}, \quad (3)$$

where W^R is the strain energy parameter, ε^R is the strain parameter, S is the internal damage variable (typically represented by modulus), σ is the stress parameter, and α = material constant in the undamaged state.

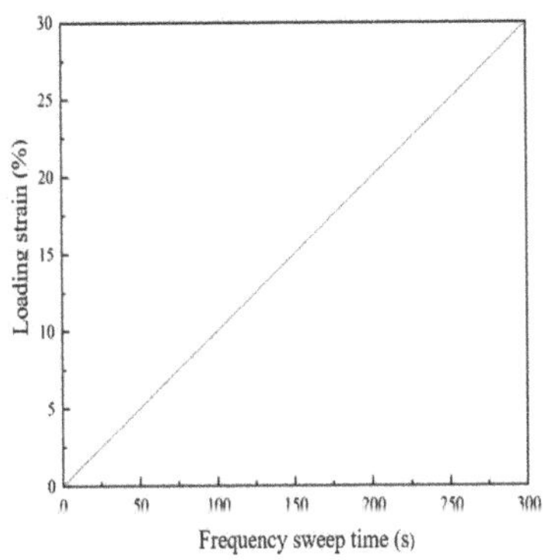

Figure 4. Loading protocol.

To apply the Viscoelastic Continuum Damage (VECD) theory to asphalt, Equation (4) is used in place of Equation (2):

$$W = \pi \cdot I_D \cdot \gamma_0^2 \cdot |G^*| \cdot \sin\delta, \quad (4)$$

where I_D is the dynamic shear modulus in the undamaged state, γ_0 is the strain amplitude (%), $|G^*|$ is the complex shear modulus, and δ is the phase angle.

The damage parameter D is quantified by linking the modulus and damage through the following equation:

$$D(t) \cong \sum_{i=1}^{N} \left[\pi\gamma_0^2(|G^*|\sin\delta_{i-1} - |G^*|\sin\delta_i)\right]^{\frac{\alpha}{1+\alpha}} (t_i - t_i)^{\frac{1}{1+\alpha}} \quad (5)$$

where I_D is the dynamic shear modulus in the undamaged state, γ_0 is the strain amplitude (%), $|G^*|$ is the complex shear modulus, and δ is the phase angle.

The damage parameter D is quantified by linking the modulus and damage through the following equation:

$$G' = |G^*| \times \cos\delta, \quad (6)$$

$$G'' = |G^*| \times \sin\delta, \quad (7)$$

By taking the logarithm of the storage modulus $G\prime$ and angular frequency ω, the linear fit in Equation (8) yields the parameter m:

$$logG' = m(log\omega) + b, \quad (8)$$

The rheological parameter α is then calculated as

$$\alpha = \frac{1}{m}, \tag{9}$$

To establish the relationship between the internal state of the material and the damage parameter, the following fitting model is used:

$$C(t) = C_0 - C_1[D(t)]^{C_2}, \tag{10}$$

where $C_0 = 1$, and the fitting parameters C_1 and C_2 are derived from

$$lg[C_0 - C(t)] = lgC_1 + C_2 lg[D(t)], \tag{11}$$

The fatigue failure criterion evaluates asphalt fatigue damage and predicts fatigue life under varying loading conditions. According to AASHTO TP 101-12, fatigue failure is defined when the initial $|G^*|\sin\delta$ drops to 35% (i.e., $C = 0.65$). However, this criterion treats all asphalt samples uniformly, ignoring material variability and leading to less accurate fatigue life predictions.

In contrast, AASHTOTP 101-14 uses the C value corresponding to the peak stress in the stress–strain curve as the fatigue failure point. Zhou et al. [28] validated both criteria, demonstrating that the peak stress-based C value better reflects true fatigue failure with lower variability. Therefore, this study adopts the peak stress-based C value for fatigue life calculation. The fatigue damage failure criterion (D_f) is expressed as

$$D_f = \left(\frac{C_0 - C_{peak}}{C_1}\right)^{\frac{1}{C2}}, \tag{12}$$

where C_{peak} is the $C(t)$ value at peak stress.

The fatigue equation parameters A and B are calculated as

$$A = \frac{f\left(D_f\right)^k}{k(\pi C_1 C_2)^{\alpha}}, B = 2\alpha, \tag{13}$$

The final fatigue life equation, used to predict asphalt fatigue life at any strain level, is

$$N = A(\gamma_{max})^{-B}, \tag{14}$$

where f = 10 Hz, and $k = 1 + (1 - C_2)\alpha$.

2.5.3. Fatigue–Healing–Fatigue Test

Xie et al. [29] proposed the linear amplitude sweep-based healing test (LASH) to quantify the self-healing properties of asphalt. In this study, the LASH test was conducted by introducing a rest period into the standard LAS test. The self-healing performance was evaluated by comparing the damage variable D before and after the rest period.

The key aspects of the LASH test include the selection of the damage level and the healing time. For the asphalt samples subjected to coupled aging in this study, the LAS test was first performed to determine the damage parameter D_f at fatigue failure, defined by the peak stress. A damage level of 50% D_f was selected as the threshold for initiating the healing process, and the corresponding strain level was calculated.

A new asphalt specimen was then subjected to the LAS test at the standard loading rate. When the strain reached the predetermined level (corresponding to 50% D_f), the loading was stopped, and the specimen was allowed to rest for one hour at a constant temperature of 25 °C. The second loading phase began with the same strain amplitude and loading rate as the end of the first phase. The loading protocol is illustrated in Figure 5.

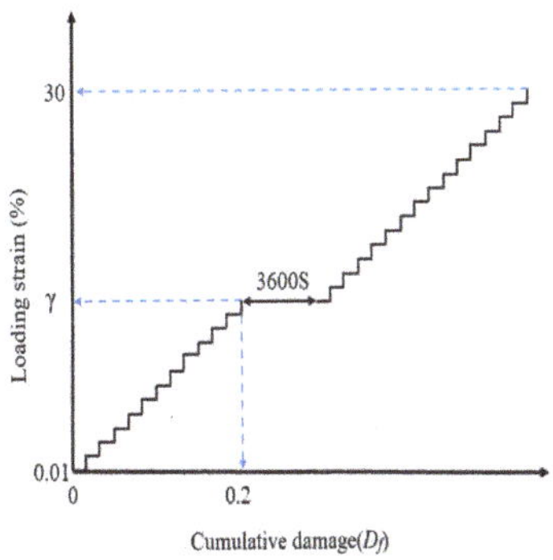

Figure 5. Linear amplitude sweep-based healing test protocol.

3. Results and Discussion

3.1. Linear Viscoelastic Analysis (Continued)

The dynamic shear modulus (G^*) is a critical parameter for characterizing the viscous and elastic behavior of asphalt materials under dynamic loading. An increase in G^* at low frequencies (high temperatures) enhances the asphalt's resistance to high-temperature deformation. Conversely, an increase in G^* at high frequencies (low temperatures) leads to hardening and embrittlement, resulting in poor low-temperature performance. The master curves of the dynamic shear modulus for SBS-T-modified asphalt under different coupled aging conditions are shown in Figure 6.

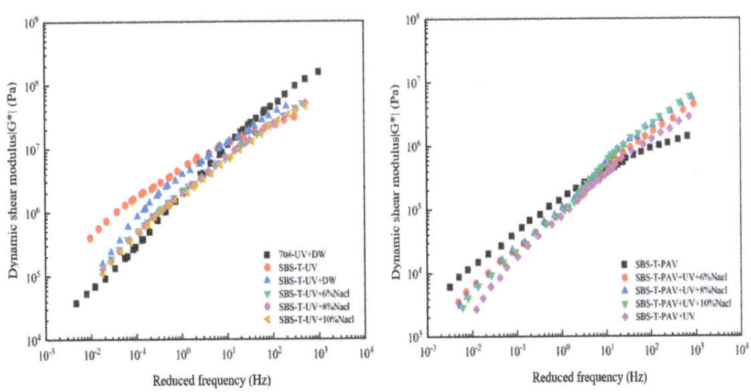

(**a**) (**b**)

Figure 6. (**a**) UV–salt solution aging, and (**b**) UV–pressure aging and UV–pressure–salt solution aging.

As shown in Figure 7, in the low-frequency (high-temperature) region, SBS-T-modified asphalt exhibits a higher dynamic shear modulus (G^*) compared to base asphalt. In contrast, in the high-frequency (low-temperature) region, the base asphalt demonstrates a higher G^* than SBS-T-modified asphalt. The significant enhancement in modulus across the entire range indicates that base asphalt is more sensitive to temperature variations, leading to poor performance at both low and high temperatures. Specifically, it is prone to cracking at low temperatures and deformation at high temperatures. On the other hand, SBS-T-modified asphalt exhibits superior high-temperature deformation resistance and low-temperature crack resistance.

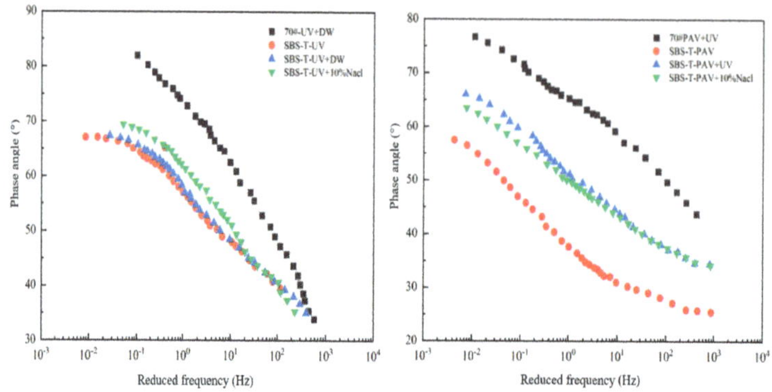

(**a**) (**b**)

Figure 7. (**a**) UVsalt solution aging, and (**b**) UVpressure aging and UVpressuresalt solution aging.

In the low-frequency region (Figure 6a), the dynamic shear modulus of SBS-T-modified asphalt under different aging conditions follows the order of UV aging alone > UV–deionized water coupling > UV–salt solution coupling. The UV–salt solution coupling significantly reduces G^* compared to UV aging alone, indicating that the combined effect of UV and salt solution deteriorates the deformation resistance of asphalt. In contrast, as shown in Figure 6b, the dynamic modulus of SBS-T-modified asphalt under UV–pressure–salt solution coupling is higher than that under UV–pressure coupling, suggesting that the addition of salt solution improves the deformation resistance when combined with pressure and UV aging.

In the high-frequency region, SBS-T-modified asphalt subjected to UV aging alone exhibits a lower G^*, while UV–deionized water and UV–salt solution coupling result in higher G^*. This indicates that coupling UV aging with these factors hardens and embrittles the asphalt, reducing its low-temperature crack resistance. However, the dynamic modulus of asphalt decreases after UV–salt solution aging, suggesting that the salt solution improves the low-temperature performance of SBS. Conversely, the dynamic modulus increases after UV–pressure–salt solution aging, indicating that the salt solution weakens the crack resistance under these conditions.

The phase angle (δ) represents the ratio of viscous-to-elastic response in asphalt and reflects the material's viscoelastic behavior. A higher δ indicates a greater viscous component, leading to poorer deformation resistance at high temperatures. Conversely, a lower δ suggests a stronger elastic response, enhancing the material's ability to resist deformation. The master curves of δ for asphalt under different aging conditions are shown in Figure 7.

As shown in Figure 7, the phase angle (δ) master curves of all asphalt samples generally decrease with increasing frequency. This trend occurs because, as the frequency increases, the loading duration shortens, causing the asphalt's deformation to transition from viscous to elastic behavior, thereby reducing the phase angle. Under the same aging conditions, the phase angle curves of base asphalt are consistently above those of SBS-T-modified asphalt, indicating that SBS-T modification enhances the deformation resistance of asphalt.

As aging factors become more complex, the phase angle increases, suggesting that coupled aging introduces more viscous components into the asphalt, reducing its deformation resistance. Specifically, compared to UV aging alone, the UV–salt solution coupling results in a larger phase angle for SBS-T-modified asphalt, indicating that the salt solution accelerates aging and further hardens the asphalt. This hardening increases the viscous behavior of asphalt at high temperatures. The underlying mechanism may involve the intense UV aging reaction, where the introduction of a salt environment disrupts the colloidal structure balance within the asphalt, accelerating the breakdown of chemical bonds and their reaction with oxygen.

In contrast, when pressure aging is included in the coupling, the phase angle decreases significantly, indicating an increase in elastic components and enhanced deformation resistance. Compared to UV–pressure coupling, the UV–pressure–salt solution coupling results in a smaller phase angle, suggesting that the salt environment promotes a higher elastic response during pressure aging, further improving the asphalt's deformation resistance.

3.2. Fatigue Performance Analysis

The peak stress in the stress–strain curve was used as the fatigue failure criterion in the LAS test. Based on the VECD fatigue criterion, the stress–strain curves of asphalt under different aging conditions at 25 °C were analyzed, as shown in Figure 8.

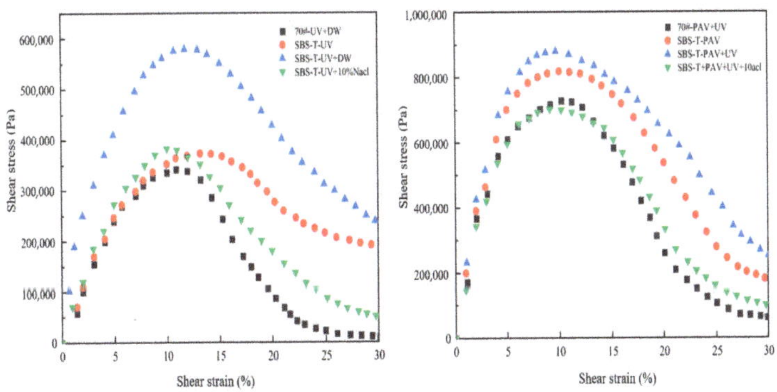

(a) (b)

Figure 8. (**a**) UVsalt solution aging, and (**b**) UVpressure aging and UVpressuresalt solution aging.

From Figure 8, it can be observed that the peak stress of the asphalt material can be determined by analyzing the stress–strain curves, providing insights into the strain-dependent performance of the material. Under different coupled aging conditions, the asphalt exhibits peak stress, followed by a decrease in stress after the peak strain, indicating the occurrence of fatigue damage.

All asphalt samples display a plateau region at the peak stress, with a wider plateau indicating stronger resistance to loading. Under the same aging conditions, SBS-T-modified asphalt exhibits a broader peak stress region compared to base asphalt, and a plateau region emerges as the strain increases. This suggests that SBS-T-modified asphalt is less sensitive to strain and has enhanced deformation resistance, reflecting the influence of the SBS-T modifier on its viscoelastic properties. In the peak stress region, a decrease in stress signifies the initiation of fatigue damage. Compared to base asphalt, SBS-T-modified asphalt shows a slower stress reduction, indicating that strain variations have a lesser impact on stress.

As aging factors become more complex, the peak stress of SBS-T-modified asphalt increases, while the peak strain decreases. The peak stress of SBS-T-modified asphalt after UV–deionized water coupling is higher than that after UV–salt solution coupling. Similarly, the peak stress after UV–pressure coupling is greater than that after UV–pressure–salt solution coupling. The introduction of a salt environment significantly reduces the peak stress, indicating that the salt solution diminishes the deformation resistance and increases the strain dependency of the asphalt, adversely affecting the durability of asphalt pavements. Compared to pressure aging alone, UV–pressure coupling increases the peak stress of SBS-T-modified asphalt, suggesting that UV aging influences the deformation resistance, making the asphalt more susceptible to physicochemical changes under UV exposure.

The DCC, derived from the VECD theory and the peak stress fatigue criterion, reflects the relationship between pseudo-stiffness modulus (C) and cumulative damage variable (*D*). The DCC curves of aged asphalt are shown in Figure 9.

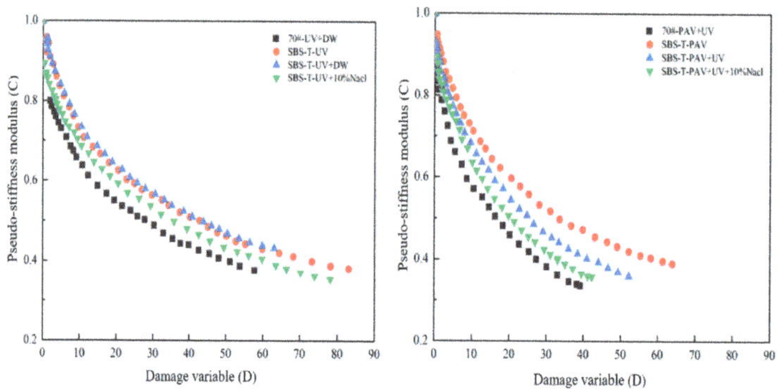

(a) (b)

Figure 9. (**a**) UVsalt solution aging, and (**b**) UVpressure aging and UVpressuresalt solution aging.

Each DCC exhibits distinct C_f and D_f values, where a higher D_f typically indicates better material durability. The endpoints of the curves in Figure 9, determined based on the peak stress fatigue criterion, represent the fundamental fatigue characteristics of the material. As shown in Figure 10, the endpoints of the DCC curves vary under different aging conditions, reflecting changes in the physical and chemical structures of the asphalt due to aging.

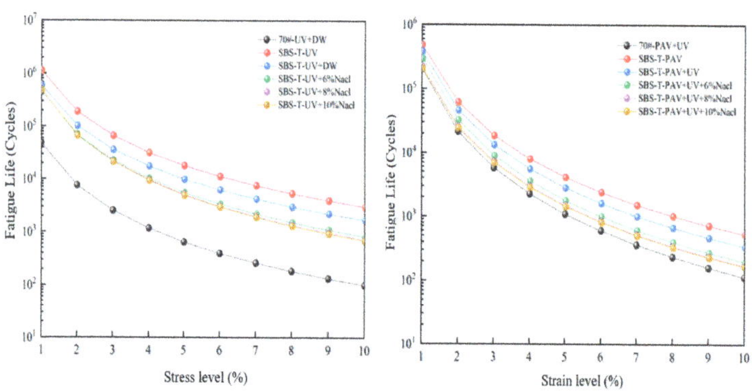

(a) (b)

Figure 10. (**a**) UV–salt solution aging, and (**b**) UV–pressure aging and UV–pressure–salt solution aging.

As loading continues, damage accumulates in the asphalt, and the pseudo-stiffness modulus (C) decreases from C = 1 to the defined fatigue damage point. The decreasing rate of the curve slows over time, indicating that the sensitivity of asphalt fatigue damage to loading gradually diminishes. Under the same aging conditions, base asphalt exhibits the smallest damage threshold, while SBS-T-modified asphalt shows higher damage thresholds, demonstrating that SBS-T modification enhances the fatigue durability of asphalt.

After coupled aging, the endpoints of the DCC curves decrease to varying degrees. The DCC curve of SBS-T-modified asphalt under UV–salt solution coupling lies below that of UV aging alone, indicating faster damage accumulation. This suggests that, at the same damage level, asphalt not exposed to a salt environment maintains better structural integrity, while the salt environment reduces fatigue performance.

Analyzing the four asphalt samples subjected to pressure aging, the DCC curves are ranked as follows: SBS-T–PAV > SBS-T–PAV + UV > SBS-T–PAV + UV + 10% NaCl > base asphalt–PAV + UV. This ranking demonstrates that, as aging factors become more complex, the performance degradation of SBS-T-modified asphalt becomes more severe, with the most significant deterioration occurring under the three-factor coupling (pressure, UV, and salt solution).

Using Equation (14), the fatigue life of asphalt at any strain level can be predicted. Figure 10 shows the predicted fatigue life of asphalt under different aging conditions at strain levels of 1–10%.

As illustrated in Figure 10, asphalt fatigue life exhibits an inverse correlation with the strain magnitude, declining progressively as the strain increases. The fatigue life of SBS-T-modified asphalt subjected to UV–salt solution coupled aging is notably shorter than that under UV aging alone, highlighting the detrimental synergistic effects of saline environments on fatigue resistance. This observation aligns with field evidence showing heightened susceptibility to fatigue cracking in salt-exposed pavements under sustained traffic loads.

While SBS-T-modified asphalt demonstrates superior fatigue life compared to base asphalt after UV–salt solution coupling, the inclusion of pressure aging diminishes this advantage. Specifically, pressure aging disrupts the colloidal structure of SBS-T-modified asphalt, reducing cohesion and accelerating crack initiation. Post-pressure–UV coupling, the fatigue life of SBS-T-modified asphalt further decreases, underscoring UV aging's incremental degradation effect. The most severe fatigue life reduction (lowest among all the conditions) occurs under triaxial coupling (UV–pressure–salt), emphasizing the compounding influence of multi-factor aging.

Existing studies correlate the maximum strain thresholds (γ_{max}) with pavement layer thickness: $\gamma_{max} = 2.5\%$ for thick layers and 5% for thin layers [30]. At $\gamma_{max} = 5\%$, asphalt binder fatigue life (N_f) strongly correlates with mixture-level fatigue performance [31]. This study calculated N_f for two strain levels ($\gamma_{max} = 2.5\%$ and $\gamma_{max} = 5\%$) across aging conditions, with the results summarized in Figure 11.

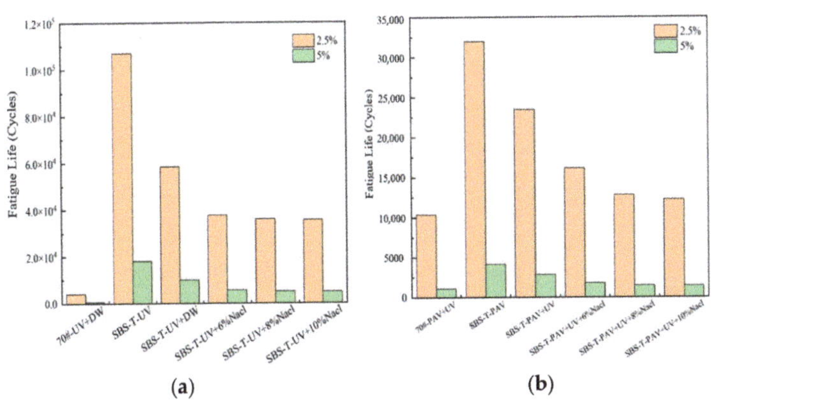

Figure 11. (**a**) UV–salt solution aging, and (**b**) UV–pressure aging and UV–pressure–salt solution aging.

As illustrated in Figure 11, the addition of the SBS-T modifier enhances the fatigue life of asphalt. At both the 2.5% and 5% strain levels, the trends in fatigue life under different aging conditions are consistent. The fatigue life of SBS-T-modified asphalt after

UV–salt solution coupling is shorter than that after UV aging alone, indicating that both UV radiation and the salt environment negatively impact the service life of asphalt pavements. When UV and salt solution coupling is applied, increasing the salt concentration has a minimal effect on fatigue life, suggesting that salt concentration plays a minor role in influencing fatigue performance under these conditions.

The fatigue life of SBS-T-modified asphalt after pressure–UV–salt solution coupling is the shortest among all the coupled aging conditions. This demonstrates that the combined action of these three factors disrupts the internal colloidal structure of the asphalt, destabilizing its matrix and reducing its modulus, thereby significantly impairing fatigue performance. In real-world pavement applications, the pressure–UV–salt solution coupling makes the pavement more prone to cracking, severely compromising its service life.

At the two predefined strain levels (2.5% and 5%), the fatigue life of SBS-T-modified asphalt is comparable to that of the base asphalt. At the lower strain level (2.5%), complex aging factors reduce the fatigue resistance of asphalt, and the fatigue life decreases to varying degrees under salt exposure. At the higher strain level (5%), SBS-T-modified asphalt demonstrates some advantages in fatigue resistance, but its resistance to salt-induced degradation remains insufficient.

3.3. Self-Healing Performance Analysis

In this study, the autogenous healing efficiency was quantified through Equation (15) to evaluate asphalt's self-restorative capacity during intermittent loading cycles:

$$\%H_S = \frac{D_1 - D_2}{D_1} \times 100\%, \tag{15}$$

where D_1 is the damage variable before the rest period, and D_2 is the damage variable after the rest period.

The self-healing index, calculated using Equation (15), reflects the relationship between pseudo-stiffness and damage intensity. Elevated healing indices directly correlate with the enhanced intrinsic healing potential of the asphalt binder. Figure 12 presents a comparative evaluation of autogenous recovery efficiencies across specimens subjected to thermal–oxidative aging, UV-induced degradation, and their synergistic effects.

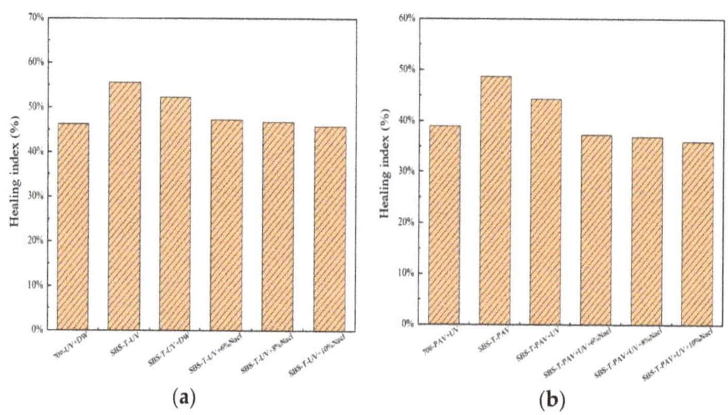

Figure 12. (**a**) UV–salt solution aging, and (**b**) UV–pressure aging and UV–pressure–salt solution aging.

As illustrated in Figure 12, under the same aging conditions, SBS-T-modified asphalt exhibits a higher healing index ($\%H_S$) compared to base asphalt, indicating that the addition of the SBS-T modifier enhances the self-healing capability of asphalt. This improvement is attributed to the elastic recovery properties of the SBS-T modifier's molecular chains.

Compared to UV aging alone, the healing index of SBS-T-modified asphalt significantly decreases after UV–salt solution coupling, demonstrating that the combined effect of these two aging factors reduces the healing capability of asphalt. When long-term aging factors are introduced, the self-healing efficiency of asphalt under pressure–UV–salt solution coupling is the lowest. This is largely due to UV-induced hardening, which degrades the mechanical properties of asphalt and hinders its healing process. Additionally, long-term aging causes significant changes in the chemical structure of asphalt molecules, slowing down their diffusion rate. These findings suggest that more complex aging environments lead to lower self-healing efficiency in SBS-T-modified asphalt.

Furthermore, analyzing the presence of salt solution in the aging conditions reveals that the healing index of SBS-T-modified asphalt decreases more significantly under salt solution coupling compared to deionized water exposure. This indicates that deionized water has a minimal impact on the healing behavior of asphalt, while salt solutions significantly deteriorate its self-healing capability. The deterioration is attributed to the adsorption of salt ions at the interface between asphalt molecules and the salt solution, which hinders the diffusion of asphalt molecules at the healing interface. Therefore, in real-world applications, particularly in coastal areas or during winter when deicing salts are used, the combined effects of UV radiation and long-term aging further reduce the self-healing ability and durability of asphalt pavements.

The inclusion of a rest period allows partial recovery of the asphalt's modulus, leading to an improvement in fatigue life. Based on the VECD theory, the fatigue life of asphalt before and after self-healing at a shear strain of 5% was calculated, as illustrated in Figure 13.

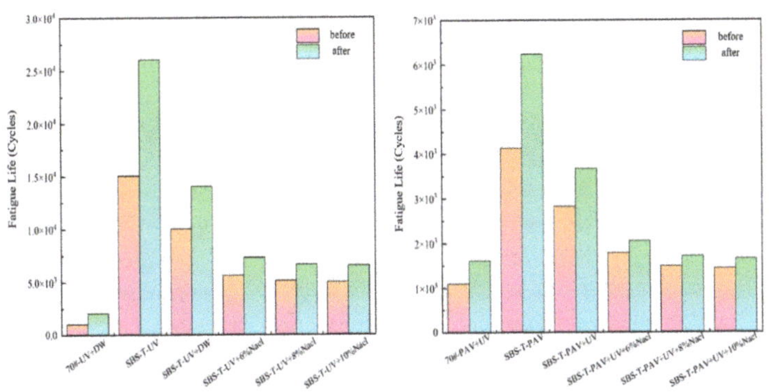

(a) (b)

Figure 13. (**a**) UV–salt solution aging, and (**b**) UV–pressure aging and UV–pressure–salt solution aging.

As illustrated in Figure 13, the fatigue life of asphalt increases significantly after the rest period. SBS-T-modified asphalt exhibits the most pronounced improvement in fatigue life after UV aging and pressure aging, while the growth rate of fatigue life decreases under multi-factor coupling conditions. This indicates that the combined effects of multiple aging factors hinder the recovery of the dynamic modulus during the rest period. Among the three coupled aging conditions, the inclusion of pressure aging has the greatest impact on the improvement of fatigue life before and after healing. This is attributed to the increased formation of hard components during long-term aging, which impedes the healing of internal microcracks and reduces the asphalt's ability to withstand secondary loading.

Under aging conditions coupled with a salt environment, SBS-T-modified asphalt shows the lowest fatigue life and the smallest increase in fatigue life before and after healing. This demonstrates that salt solution erosion has a significant negative impact on the fatigue life of asphalt. Compared to previous analyses, when self-healing behavior is considered, all coupled aging conditions reduce the fatigue life of asphalt, with the most severe deterioration observed under UV–pressure–salt solution coupling.

4. Conclusions

This study investigated the effects of coupled aging factors—UV–salt solution, UV–pressure, and UV–pressure–salt solution—on the fatigue damage evolution and self-healing performance of asphalt using a dynamic shear rheometer and the Viscoelastic Continuum Damage (VECD) theory. The fatigue life of asphalt under different aging conditions was quantified, and the impact of coupled aging on self-healing performance was analyzed by evaluating healing indices and fatigue life changes. The main conclusions are as follows:

(1) The salt environment under UV–salt solution coupling makes SBS-T-modified asphalt more prone to deformation, reducing its high-temperature deformation resistance but improving its low-temperature crack resistance. Under UV–pressure–salt solution coupling, the participation of pressure aging enhances the high-temperature performance of asphalt but negatively affects its low-temperature performance.

(2) As aging factors become more complex, the peak stress of SBS-T-modified asphalt increases, while the peak strain decreases. The inclusion of pressure aging significantly accelerates the rate of peak stress reduction. Both UV–salt solution and UV–pressure–salt solution coupling reduce the peak stress of SBS-T-modified asphalt, indicating that the salt environment deteriorates the deformation resistance of asphalt.

(3) The pseudo-stiffness modulus of SBS-T-modified asphalt after UV–salt solution and UV–pressure–salt solution coupling is lower than that under single-factor aging, demonstrating that coupled aging reduces the deformation resistance. Under conditions involving pressure aging, the damage characteristic curve (DCC) shows a faster decline, indicating that pressure aging accelerates fatigue and weakens fatigue durability. The fatigue life of asphalt significantly decreases after coupling with salt solution and other aging factors, highlighting the adverse impact of salt solution on fatigue performance.

(4) Based on the VECD model and healing index, the self-healing performance of SBS-T-modified asphalt decreases after all three coupled aging conditions. The most significant reduction in self-healing index occurs under UV–pressure–salt solution coupling, where the self-healing index decreased by approximately 11% compared to UV–pressure aging alone. Furthermore, the post-healing fatigue life under this coupled aging condition was the lowest, showing a 70% reduction relative to UV–pressure aging. These results demonstrate that the combined effect of these three factors has the most pronounced impact on the self-healing performance of asphalt.

Author Contributions: Conceptualization, J.Y. and R.L.; methodology, R.L.; software, Y.W.; validation, Y.W. and J.F.; formal analysis, Y.W.; investigation, J.F.; resources, J.F.; data curation, R.L.; writing—original draft preparation, J.F.; writing—review and editing, J.F.; visualization, Y.M.; supervision, R.L.; project administration, Y.M. and G.T.; funding acquisition, Y.M. and G.T. All authors have read and agreed to the published version of the manuscript.

Funding: This research was funded by [Research on Key Technologies of Low-Carbon Durable High Rutting Resistance Asphalt Surface Layer Structure and Materials] grant number [HNJS-2024-K28].

Institutional Review Board Statement: Not applicable.

Informed Consent Statement: Not applicable.

Data Availability Statement: The original contributions presented in this study are included in the article. Further inquiries can be directed to the corresponding author(s).

Conflicts of Interest: Author Yang Ming was employed by the company Henan Transport Investment Group Co., Ltd. Author Guoqi Tang was employed by the company Guolu Gaoke Engineering Technology Institute Co., Ltd. The remaining authors declare that the research was conducted in the absence of any commercial or financial relationships that could be construed as a potential conflict of interest.

References

1. Zhuang, Y.; Yue, J.; Men, B.; Tang, G.; Wang, R. Experimental Study on Mechanism, Aging and Fatigue Performance of Warm Mixing Speed Melting SBS Modified Asphalt Binders. *Coatings* **2023**, *13*, 0311. [CrossRef]
2. Xu, Y. Study on Influencing Factors and Correlation of Asphalt Fatigue and Self-Healing Properties. Master's Thesis, Wuhan University of Technology, Wuhan, China, 2018.
3. Zhang, R.; Yu, H. Influence of Different Aging Environments on Rheological Behavior and Structural Properties of Rubber Asphalt. *Materials* **2020**, *13*, 33–43. [CrossRef] [PubMed]
4. Sun, S.; Wang, Y. Application of Ultraviolet Absorbers in Asphalt Mixtures. *Highw. Eng.* **2012**, *37*, 109–114.
5. Wang, J. Study on Ageing Mechanism and Rheologic Behavior of Asphalt Under Simulated Ultraviolet Environment. Ph.D. Thesis, Harbin Institute of Technology, Harbin, China, 2008.
6. Zhou, Z.; Jiang, T.; Jiang, W. Evaluation of Water Damage Resistance of Asphalt Mixtures in High-Temperature and Humid Regions. *J. Chang. Univ. Sci. Technol. (Nat. Sci.)* **2013**, *10*, 8–14.
7. Xiong, R.; Chu, C.; Qiao, N.; Wang, L.; Yang, F.; Sheng, Y.; Guan, B.; Niu, D.; Geng, J.; Chen, H. Performance Evaluation of Asphalt Mixture Exposed to Dynamic Water and Chlorine Salt Erosion. *Constr. Build. Mater.* **2019**, *201*, 121–126. [CrossRef]
8. Zhang, Q.; Huang, Z. Investigation of the Microcharacteristics of Asphalt Mastics under Dry-Wet and Freeze-Thaw Cycles in a Coastal Salt Environment. *Materials* **2019**, *12*, 2627. [CrossRef] [PubMed]
9. Huang, X.Y.; Sha, A.M.; Jiang, W.; Wang, Y. Effect and Mechanism of Salt Erosion on Performance of Bitumen and Asphalt Mixtures. *J. Chang'an Univ. (Nat. Sci. Ed.)* **2017**, 33–38, 46. [CrossRef]
10. Zhang, Q.; Wu, D.; Zhang, X.; Chang, K.; Wang, Y. Effect of Organic Deicing Agents on Asphalt Rheology and Analysis of the Mechanism. *Constr. Build. Mater.* **2021**, *284*, 122649. [CrossRef]
11. Polacco, G.; Stastna, J.; Biondi, L. Relation between Polymer Architecture and Nonlinear Viscoelastic Behavior of Modified Asphalt. *Eur. Polym. J.* **2006**, *42*, 446–457. [CrossRef]
12. Xia, X.; Tang, D.; Wang, P.; Zhao, Y. Synergistic effect of polymer-modified rejuvenator and fresh asphalt on road performance of hot in-place recycling styrene–butadiene–styrene-modified asphalt mixtures. *Polym Int.* **2025**. [CrossRef]
13. Zhuang, H.; Zhong, H.; Wang, J.; Xian, L.; Li, D. Comparative Analysis of High-Temperature Performance of SBS-Modified Asphalt and Its Mixtures. *Highw. Eng.* **2023**, *48*, 116–122.
14. Mao, S.; Xiong, L.; Lin, X.; Yang, K. Study on Evaluation Indexes of Storage Stability of SBS-Modified Asphalt. *Pet. Asph.* **2012**, *26*, 11–15.
15. Sun, D.; Lü, W. Study on Thermal Storage Stability of SBS-Modified Asphalt. *J. Build. Mater.* **2006**, *9*, 671–674.
16. Wang, H. Directly Added SBS Modifier and Its Preparation Method. CN103509358A, 15 January 2014.
17. Wang, Z.; Liu, C.; Zhao, Y.; Wen, L.; Li, K.; Liu, B.; Hu, F. Study on Road Performance of Directly Added SBS Modifier. *J. Highw. Transp. Res. Dev. (Appl. Technol. Ed.)* **2016**, *12*, 90–92.
18. Li, R.; Wang, Y.; Zhu, W.; Chen, Y.; Yue, J. Research on the Rheological Performance of Fast-Melting SBS-Modified Asphalt under Complex Environmental Factors. *Coatings* **2024**, *14*, 1241. [CrossRef]
19. Wang, Y. Application of Dry-Process SBS-Modified Asphalt in Municipal Roads. *Build. Superv. Test. Cost* **2022**, *15*, 9–12, 20.
20. Feng, Z.; Yu, J.; Zhang, H.; Kuang, D.; Xue, L. Effect of Ultraviolet Aging on Rheology, Chemistry, and Morphology of Ultraviolet Absorber-Modified Bitumen. *Mater. Struct.* **2013**, *46*, 1123–1132. [CrossRef]
21. Zhang, C.; Yu, J.; Feng, K.; Xue, L.; Xie, D. Synthesis and Characterization of Triethoxyvinylsilane Surface-Modified Layered Double Hydroxides and Application in Improving UV Aging Resistance of Bitumen. *Appl. Clay Sci.* **2016**, *120*, 1–8. [CrossRef]
22. Dong, Y.; Tang, G.; Hou, Y.; Wang, H.; Tian, J.; Wei, Y. Study on aging kinetics of dry-modified asphalt based on complex modulus. *Hunan J. Transp. Sci. Technol.* **2022**, *48*, 20–22.
23. Schwarzl, F.R.; Staverman, A.J. Time-Temperature Dependence of Linear Viscoelastic Behavior. *J. Appl. Phys.* **1952**, *23*, 838–843. [CrossRef]
24. Findley, W.N.; Lai, J. A Modified Superposition Principle Applied to Creep of Nonlinear Viscoelastic Material under Abrupt Changes in State of Combined Stress. *Trans. Soc. Rheol.* **1967**, *11*, 361–380. [CrossRef]
25. Tan, Y.; Shao, X.; Zhang, X. Research on Evaluation Method of Low-Temperature Performance of Asphalt Based on Low-Temperature Rheological Properties. *China J. Highw. Transp.* **2002**, 3–7. [CrossRef]

26. Yue, M.; Yue, J.; Wang, R.; Xiong, Y. Evaluating the fatigue characteristics and healing potential of asphalt binder modified with Sasobit (R) and polymers using linear amplitude sweep test. *Constr. Build. Mater.* **2021**, *289*, 123054. [CrossRef]
27. Kang, X.; Wang, R.; Yue, J.; An, X.; Tang, G. Investigation of the adhesion characteristics of a novel fast-melting SBS-based modifier to asphalt-aggregate systems based on a multiscale approach. *J. Mater. Civ. Eng. (ASCE)* **2023**, *35*, 04023435. [CrossRef]
28. Zhou, S.; Shi, J.; Zhang, R.; Zhang, X.; Yuan, Y. Evaluation of Asphalt Fatigue Performance Based on Linear Amplitude Sweep Test. *Pet. Asph.* **2018**, *32*, 27–33.
29. Xie, W.; Castorena, C.; Wang, C.; Kim, Y.R. A Framework to Characterize the Healing Potential of Asphalt Binder Using the Linear Amplitude Sweep Test. *Constr. Build. Mater.* **2017**, *154*, 771–779. [CrossRef]
30. Wang, R.; Xiong, Y.; Yue, M.; Hao, M.; Yue, J. Investigating the Effectiveness of Carbon Nanomaterials on Asphalt Binders from Hot Storage Stability, Thermodynamics, and Mechanism Perspectives. *J. Clean. Prod.* **2020**, *276*, 124180. [CrossRef]
31. Wang, C. Rheological Study on Road Performance of Asphalt Binder. Ph.D. Thesis, Beijing University of Technology, Beijing, China, 2015.

materials

MDPI

Article

Investigation of Rutting and Aging Performance of Gap-Graded Rubberized Asphalt Mixtures

Marek Pszczola * and Bohdan Dolzycki

Faculty of Civil and Environmental Engineering, Gdansk University of Technology, 80-233 Gdansk, Poland; bohdan.dolzycki@pg.edu.pl
* Correspondence: marek.pszczola@pg.edu.pl; Tel.: +48-58-347-27-82

Abstract: Gap-graded asphalt mixtures like stone mastic asphalt (SMA), porous asphalt (PA), and asphalt mixtures for very thin layers (fr. Béton Bitumineuse Très Mince—BBTM) are usually made with the use of SBS (styrene-butadiene-styrene) polymer-modified bitumen. This is a binder that allows one to achieve the required parameters, but at the same time, its use increases the costs of making pavement layers. An alternative to polymer-modified bitumen (SBS) is rubber-modified bitumen. The research presented in this publication includes an assessment of the resistance to permanent deformation and susceptibility to aging of SMA and porous asphalt (PA) mixtures containing both SBS polymer-modified bitumen and rubber-modified bitumen, where the modification process was carried out directly in the refinery. The laboratory tests of resistance to deformation were assessed based on the rutting test and on the assessment of the dynamic modulus (SPT). The changes in the tested asphalt mixtures after aging in laboratory conditions were assessed based on the changes in the stiffness modulus (IT-CY) and the changes in the indirect tensile strength (ITS) after the short-term and long-term aging processes. The presented research results clearly show that the use of rubber-modified bitumen produced in industrial conditions (i.e., in a refinery) allows one to obtain gap-graded mixtures that are as resistant to permanent deformation as mixtures containing SBS polymer-modified bitumen. Similar conclusions resulted from the study of susceptibility to aging. Changes after aging for both types of asphalt mixtures were at a similar level. The presented results clearly indicate that, in the case of gap-graded mixtures such as SMA- and PA-type mixtures, they meet the rutting and aging expectations when either expensive modified bitumen or a cheaper, more environmentally friendly alternative (rubber-modified bitumen) is used.

Keywords: gap-graded asphalt mixtures; crumb rubber modification; polymer–rubber-modified bitumen; rutting and aging resistance

1. Introduction

Gap-graded asphalt mixtures, such as stone mastic asphalt (SMA), porous asphalt (PA), and asphalt mixtures for very thin layers (Béton Bitumineuse Très Mince—BBTM), have gained widespread use in pavement construction, especially in milder-climate regions, due to their superior durability, resistance to rutting, and ability to improve skid resistance and noise reduction [1,2]. These mixtures, characterized by the absence of intermediate aggregate sizes, form a coarse aggregate skeleton with increased stone-on-stone contact, which significantly enhances rutting resistance and load-bearing capacity [3]. According to Yue et al. [4], in their study, gap-graded asphalt mixtures improved water damage tolerance, especially in hot climates, due to their thicker asphalt film and optimized aggregate structure.

Xu et al. [5] showed that porous asphalt pavements significantly reduce noise, even during rain, by weakening the "pump suction" and "trumpet" effects that generate loud tire noises. This not only improves environmental conditions but also helps drivers stay more focused in wet-weather conditions. On the other hand, Xu et al. [5] highlighted the maintenance challenges in pavements made with use of gap-graded asphalt mixtures, where clogging of air voids over time diminishes permeability and noise reduction performance, necessitating regular surface treatments or cleaning.

Gap-graded asphalt mixtures typically incorporate polymer-modified bitumen (PmB), specifically styrene-butadiene-styrene (SBS)-modified bitumen, which enhances the mechanical performance of asphalt mixtures by improving their resistance to permanent deformation, fatigue cracking, and oxidative aging [6,7]. However, SBS-modified bitumen significantly increases material costs, raising concerns about the economic feasibility of large-scale road infrastructure projects [8].

An alternative to SBS-modified bitumen is polymer–rubber-modified bitumen (RMB), which is obtained by incorporating recycled rubber—often from waste tires—into the bitumen mixtures. This modification process promotes environmental sustainability by reducing tire waste while simultaneously improving the binder's elasticity, thermal stability, and durability [9,10]. Studies have shown that polymer–rubber-modified bitumen can achieve performance levels comparable to SBS-modified bitumen, particularly in terms of aging resistance and rutting performance [11,12]. Bitumen modification with rubber can be carried out using two primary approaches: (i) the wet process, in which rubber particles are blended into bitumen at asphalt plants, and (ii) the industrial (refinery-based) process, where polymer–rubber-modified bitumen is produced under controlled conditions before transportation to the construction site [13]. The method of production in the refinery method, which, in the literature, is also called the "terminal blend" method, ensures better homogeneity, stability, and consistency in terms of the binder's performance [14,15]. Swieczko-Zurek et al. [16] stated that, according to the mix of crumb rubber and SBS polymer, applied technology allows one to reduce the amount of added SBS polymer with respect to the standard polymer-modified bitumens. Also, the new material fulfills all requirements set for PmB bitumen according to the European standard for polymer-modified bitumen, EN 14023 [17], which is very important for the material acceptance process in road investment projects. Pszczola et al. [18] conducted research on the low-temperature properties of rubberized asphalt mixtures and proved that rubber–polymer-modified bitumen presented similar resistance to low-temperature action than SBS polymer-modified bitumen. Despite these advantages, a comprehensive evaluation of SBS-modified and rubber-modified bitumen in gap-graded mixtures such as SMA and PA mixtures in higher temperatures is still required.

The objective of this research was to assess the performance of SMA and PA mixtures containing both SBS polymer-modified bitumen and rubber-modified bitumen produced in industrial conditions. The key areas of evaluation included resistance to permanent deformation through rutting tests and dynamic modulus analysis, as well as susceptibility to aging through stiffness modulus (IT-CY) and indirect tensile strength (ITS) measurements following short-term and long-term aging. The results will provide valuable insights into the feasibility of rubber-modified bitumen as a cost-effective and environmentally sustainable alternative to SBS-modified bitumen in high-performance asphalt pavements.

2. Materials and Methods

2.1. Materials

Laboratory tests were conducted on two types of asphalt mixtures for the wearing course—stone matrix asphalt (SMA 8) and porous asphalt (PA 8). Aggregate skeleton was designed in compliance with Technical Guidelines [19] and based on European standards (EN 13108-5 [20] for SMA mixtures and EN 13108-7 [21] for PA mixtures). For each type of mixture, one optimum bitumen content was selected based on laboratory tests results for typical SBS-modified bitumen. The composition of the mixtures and types of bitumen used are presented in Table 1.

Table 1. Composition of tested asphalt mixtures.

Property	Type of Asphalt Mixture			
	SMA8	SMA8 CR	PA8	PA8 CR
Passes, mm				
11.2	100.0		100.0	
8	94.2		91.2	
5.6	41.2		13.4	
2	25.6		6.7	
0.125	11.9		4.8	
0.625	9.7		4.1	
Type of aggregate	gneiss, granodiorite, and limestone		gneiss, granodiorite, and limestone	
Bitumen content, % w/w	7.0		6.5	
Type of bitumen	45/80-55 PmB	45/80-55 CR	45/80-65 PmB	45/80-65 CR

Four types of bitumen were selected for rutting and aging performance tests, including two types of rubber–polymer-modified bitumen—45/80-55 CR and 45/80-65 CR—and, as references, two types of SBS polymer-modified bitumen—45/80-55 and 45/80-65. The standard properties of the bitumens used in this research are shown in Table 2.

Before specimen compaction, every type of asphalt mixture was subjected to short-term aging according to standard specification AASHTO R30-02 [22].

Table 2. The properties of the bitumens used in the asphalt mixtures according to European standard EN 14023.

Property:	Type of Bitumen			
	45/80-55 PmB	45/80-55 CR	45/80-65 PmB	45/80-65 CR
Penetration in 25 °C, 0.1 mm, acc. EN 1426 [23]	43	53	52	45
R&B Temperature, °C, acc. EN 1427 [24]	60	55	72	76
Dynamic viscosity, Pa·s, acc. EN 12596 [25]				
90 °C	35.321	19.058	43.728	81.833
135 °C	1.225	0.859	1.813	1.947
160 °C	0.373	0.303	0.596	0.563

2.2. Methods

2.2.1. Resistance to Rutting Test Methods

Samples for laboratory testing were compacted in steel molds with internal dimensions of 300 × 300 × 40 mm at a temperature of 145 ± 5 °C until a compaction level of 98–100% was achieved. Samples were prepared and compacted successively in such a way so as to obtain the same time between compaction and testing of samples, ranging from 6 to 8 days. Immediately before testing, the samples were conditioned at a test temperature of 60 °C for at least 4 h but no longer than 18 h.

The rutting resistance of the asphalt mixtures was tested based on the European standard EN 12697-22 [26]. All values were assessed as the average value of two test results. The resistance of the asphalt mixtures to rutting was assessed based on the value of rut depth, RD_{AIR}, formed as a result of repeated passes of a loaded wheel at a constant temperature. During the test, the rut depth measured over the rut length was recorded.

2.2.2. Dynamic Modulus Test Method in AMPT/SPT

The test was performed on cylindrical samples with a diameter of 100 mm and a height of 150 mm, which were compacted in a gyratory press at a temperature of 145 ± 5 °C until the compaction level was within the range of 98–100%. Before the test, the upper and lower surfaces of the samples were precisely cut to obtain two flat and parallel planes. The dynamic modulus test was performed based on the proposal of the American guidelines NCHRP 9-29: PP 02 [27]. This test is an improved and simplified version of the dynamic modulus test based on the AASHTO TP62 standard [28]. The dynamic modulus test consisted of axial loading of a cylindrical sample with a vertical sinusoidal force at specified frequencies and measuring deformations using three LVDT sensors mounted equidistantly on the side of the cylindrical sample at a spacing of 120°. Figure 1 shows an image of the device and the method of mounting the sample in the device.

Figure 1. Image of the Asphalt Mixture Performance Tester/Simple Performance Test (AMPT/SPT) equipment IPC Global, Boronia, Australia.

The values measured in the test were the dynamic modulus and phase angle. The load frequencies that were set by the device were as follows: 25 Hz, 20 Hz, 10 Hz, 5 Hz, 2 Hz, 1 Hz, 0.5 Hz, 0.2 Hz, 0.1 Hz, and 0.01 Hz. The test procedure consisted of placing the sample in a chamber in which a constant test temperature was maintained. Then, the sample was loaded axially. The measurement of the dynamic modulus and phase angle was performed at 3 temperatures (+4 °C, +20 °C, and +40 °C) in the full frequency range from 25 Hz to 0.01 Hz. The test at a frequency of 0.01 Hz was performed only for the highest temperature. For each temperature, 3 cylindrical specimens were tested. The dynamic modulus tests were performed to obtain the characteristics of asphalt mixtures in the form of master curves. To determine the equation of the master curve for the dynamic modulus, we used the formula provided in the NCHRP 9-29 guidelines: PP 02 [27]:

$$log|E^*| = \delta + \frac{(Max - \delta)}{1 + e^{\beta + \gamma\{\log\omega + \frac{\Delta E_a}{19.14714}[(\frac{1}{T}) - (\frac{1}{T_r})]\}}} \quad (1)$$

where

$|E^*|$—dynamic stiffness modulus, psi (1 MPa = 145.0377 psi);

Max—maximum limiting modulus of the leading curve, psi;

T_r—reference temperature, °K;

T—test temperature, °K;

ΔE_a—activation energy (treated as a curve fitting parameter);

δ, β, γ—curve fitting parameters.

To determine the reduced frequency when determining the leading curve, the Arrhenius formula was used:

$$\log f_r = \log f + \frac{\Delta E_a}{19.14714}\left(\frac{1}{T} - \frac{1}{T_r}\right) \tag{2}$$

where

f_r—reduced frequency at reference temperature, Hz;

f—load frequency at test temperature, Hz.

The reference temperature was set to 20 °C. Master curves developed for the tested asphalt mixtures are presented in Figure 2.

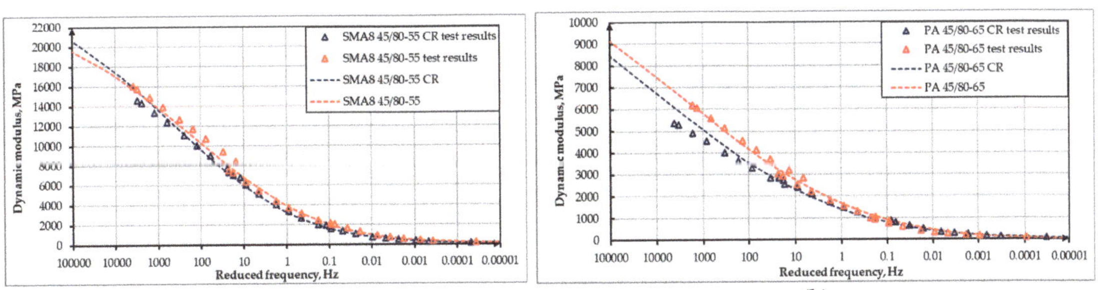

(**a**) (**b**)

Figure 2. Master curves developed for tested asphalt mixtures, where black color dots represent results of mixtures incorporating polymer–rubber-modified bitumen and red color dots represent results of mixtures incorporating SBS polymer-modified bitumen: (**a**) SMA mixture; (**b**) PA mixture.

2.2.3. Resistance to Aging Test Methods

The tests on the resistance of the analyzed asphalt mixtures to aging were carried out in accordance with the AASHTO R 30-02 procedure [22]. This procedure specifies the principles for preparing, conditioning, and testing the specimens subjected to aging:

- Short-term aging—Performed on a loose mixture, this simulates the aging of the mineral–asphalt mix during production and transport;
- Long-term aging—Performed on a compacted mixture, this simulates the operational aging of the mineral–asphalt mix in the layer after compaction.

During aging, the asphalt mixture stiffens. Long-term aging can cause it to become brittle and susceptible to cracking. Aging causes an increase in indirect tensile strength in higher summer temperatures and a decrease in low winter temperatures. The aim of this study was to compare changes in both types of asphalt mixes, differing in the type of asphalt, after the aging process.

The test was performed on cylindrical specimens with a diameter of 100 mm and a height of 63 mm, compacted by tamping (2 × 50 blow Marshall hammer) at a temperature of 145 ± 5 °C. Specimens with a diameter of D = 101 mm and a height of H = 63.5 mm were formed. The tests included unaged specimens (reference), specimens after short-term aging, and specimens after long-term aging. Each specimen group was conditioned as follows:

- Unaged specimens (reference), i.e., specimens not subjected to short-term and long-term aging—After the asphalt mixture was mixed and the mixture reached a temperature of 145 ± 5 °C, the specimens were compacted;
- Specimens after short-term aging—The asphalt mixture was mixed in a mechanical mixer at a temperature of about 145 ± 5 °C, and then the loose mixture was subjected to a short-term aging process. The mixture was spread in sheet metal forms so that the layer thickness was from 2.5 to 5.0 cm. The molds with the mixture were stored in an oven with forced air flow at a temperature of 135 ± 3 °C for 4 h ± 5 min. The mixture was cyclically mixed every hour. Then, the mixture was heated to a temperature of 145 ± 5 °C and the specimens were compacted;
- Specimens after long-term aging—Long-term aging involved storing the formed and compacted specimens after short-term aging in an oven with air flow for 5 days (120 ± 0.5 h) at a temperature of 85 ± 3 °C. The specimens were wrapped in a fine copper mesh to protect against possible deformations. An image of the samples prepared for long-term aging is shown in Figure 3.

Figure 3. Image of specimens prepared for long-term aging testing.

The assessment of the aging of asphalt mixtures consisted of comparing the test results of unaged specimens (reference) with the test results of specimens subjected to aging. The adopted aging measures were as follows:

- Change in the stiffness modulus;
- Change in the indirect tensile strength.

The tests of the stiffness modulus and indirect tensile strength were performed at a temperature of 25 °C, after a minimum of 4 h of conditioning. Each specimen was tested twice:

- For the first time, they were tested by performing a non-destructive test of the stiffness modulus following an indirect tensile scheme (IT-CY);
- for the second time, they were tested by performing a destructive test of the indirect tensile strength.

The stiffness modulus test was performed following an indirect tensile scheme (IT-CY) in accordance with Annex C of the EN 12697-26 standard [29]. The load increase time was 120 ± 4 ms, and the duration of one cycle was 3 s. The test was carried out via a controlled deformation test with a horizontal deformation of 5 μm. The test result is the average value of the stiffness modulus measured in two perpendicular planes. The indirect tensile strength tests were performed on the same samples on which the stiffness modulus test was previously performed using the indirect tensile method. This test was performed via a press with a constant displacement rate of 50 mm/min. The load was transferred through spacers that were 12 mm wide and with a radius of curvature of 50.5 mm. The

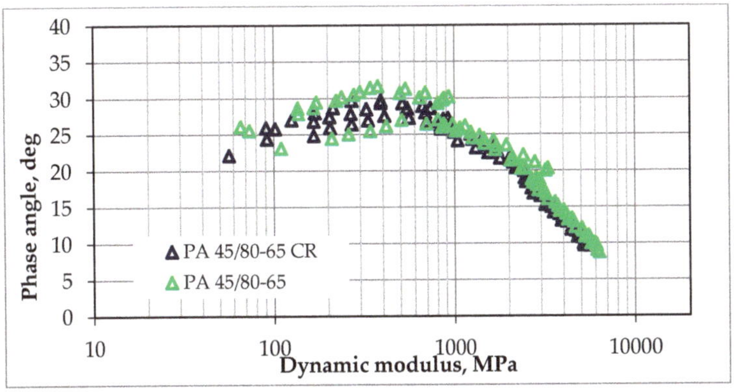

Figure 8. The results of our dynamic modulus testing of the PA mixtures according to guidelines, NCHRP 9-29 [27]; the results are presented as Black curves.

These graphs allow for the detection of nonlinearity in the behavior of asphalt mixtures and a comparison of the rheological properties of mixtures by directly presenting the results of the dynamic modulus test (without transposing the test results, as is the case when creating leading curves).

The Cole–Cole diagrams and Black curves presented do not show any significant differences in the behavior of mixtures made using SBS polymer-modified bitumen and rubber–polymer-modified bitumen (RMB).

3.3. Test Results and Analysis of the Resistance to Aging

The aging of the gap-graded asphalt mixtures was carried out according to the procedure described in Section 2.2.3. The resistance of the analyzed asphalt mixtures to aging was assessed by the following metrics:

- Index of stiffness modulus (ISM);
- Indirect tensile strength index (ITSI).

The index of stiffness modulus was defined as follows:

$$\text{ISM} = \frac{\text{mean value of stiffness modulus after aging}}{\text{mean value of stiffness modulus of reference specimens}} \cdot 100\% \tag{3}$$

The indirect tensile strength index was defined as follows:

$$\text{ITSI} = \frac{\text{mean value of indirect tensile strength after aging}}{\text{mean value of indirect tensile strength of reference specimens}} \cdot 100\% \tag{4}$$

The test results of the stiffness modulus and indirect tensile strength, as well as the indices calculated on the basis of the test results, are presented in Table 3.

Having conducted an analysis of the resistance of gap-graded asphalt mixtures to aging, we can make the following conclusions:

- In the case of the SMA mixture, the comparison of stiffness modulus indices shows that the use of rubber to modify bitumen did not affect either short-term or long-term aging;
- In the case of the porous asphalt (PA) mixture, the use of rubber to modify the bitumen significantly reduces the value of the index of stiffness modulus (ISM). The high values of the stiffness modulus obtained for the porous asphalt mixture in comparison with the SMA mixture are noteworthy. This may indicate that the PA mixture has greater susceptibility to aging, especially long-term aging;

- In the case of the SMA mixture, the comparison of indirect tensile strength indices shows that the use of rubber to modify bitumen has a beneficial effect on both short-term and long-term aging;
- In the case of the porous asphalt (PA) mixture, the addition of rubber to modified bitumen did not significantly affect the value of the indirect tensile strength index (ITSI). The values obtained regarding the indirect tensile strength indices for the porous asphalt (PA) mixture are not as high as those for the stiffness modulus indices;
- The tests validated our expectation that, with aging, the stiffness of the materials will increase, regardless of the bitumen used, especially in mixtures with higher void content values (PA 8). On the other hand, changes in the indirect tensile strength showed that, during aging, the material does not become more brittle and therefore more susceptible to cracking, regardless of the type of the mixture tested and the type of bitumen used.

Table 3. The results of our aging assessment of open-graded asphalt mixtures.

Parameter Tested:	Open-Graded Asphalt Mixture			
	SMA8 45/80-55	**SMA8 45/80-55 CR**	**PA8 45/80-65**	**PA8 45/80-65 CR**
Stiffness Modulus at 25 °C, MPa				
Reference specimens	2270	1607	641	1103
After short-term aging, MPa	2937	2075	1023	1573
After long-term aging, MPa	3150	2465	1335	2088
Index of stiffness modulus, ISM				
ISM—after short-term aging, %	129	129	159	143
ISM—after long-term aging, %	138	153	208	189
Indirect tensile strength at 25 °C, MPa				
Reference specimens	1.08	1.01	0.43	0.53
After short-term aging, MPa	1.33	1.04	0.57	0.69
After long-term aging, MPa	1.24	1.09	0.57	0.74
Indirect tensile strength index, ITSI				
ITSI—after short term aging, %	123	103	132	130
ITSI—after long-term aging, %	115	108	132	139

4. Conclusions

The presented test results regarding the rutting and aging performance of gap-graded asphalt mixtures SMA8 and PA8 with SBS polymer- and CR-modified bitumen allow us to form the following conclusions:

1. Based on the rutting test, it can be concluded that polymer-modified bitumen supplemented with rubber allows for the design of asphalt mixtures that are resistant to permanent deformations and do not differ substantially from the same mixtures containing bitumen without the addition of rubber.
2. Dynamic moduli and phase angles show a less elastic character of mixtures made using rubber–polymer-modified bitumens (lower dynamic moduli and higher phase angles at temperatures from 4 °C to 40 °C), but the differences were not significant.
3. The aging tests showed that, with aging, the stiffness of the materials increases irrespective of the type of bitumen used, especially in the more open asphalt mixture (PA 8). The increase in stiffness after aging can be more significant depending on the gap-graded asphalt mixtures' resistance to low-temperature cracking.

The research presented in this paper confirmed literature studies stating that the addition of rubber to SBS polymer bitumen does not significantly impair the properties of gap-graded asphalt mixtures in higher, summer temperatures. The PA- or SMA-type mixes made with rubber–polymer-modified bitumen behave in a similar way to those incorporating bitumen with SBS polymer modification. Laboratory tests obviously need to be verified in the field, but the results clearly show that it is possible to use cheaper rubber–polymer blended bitumens instead of more expensive SBS polymer-modified bitumens without degrading asphalt mixtures' resistance to permanent deformation.

Author Contributions: Conceptualization, M.P. and B.D.; methodology, M.P. and B.D.; formal analysis, M.P. and B.D.; writing—original draft preparation, M.P. and B.D.; writing—review and editing, M.P. and B.D. All authors have read and agreed to the published version of the manuscript.

Funding: This research received no external funding.

Institutional Review Board Statement: Not applicable.

Informed Consent Statement: Not applicable.

Data Availability Statement: The original contributions presented in this study are included in the article. Further inquiries can be directed to the corresponding author.

Acknowledgments: The authors would like to thank Mariusz Jaczewski and Cezary Szydlowski for their support in the testing of materials and analysis of results.

Conflicts of Interest: The authors declare no conflicts of interest.

Abbreviations

The following abbreviations are used in this manuscript:

GGAM	Gap-graded asphalt mixture
SMA	Stone mastic asphalt
PA	Porous asphalt
BBTM	Béton Bitumineuse Très Mince
SBS	Styrene-butadiene-styrene
PMB	Polymer-modified bitumen
RMB	Rubber-modified bitumen
AMPT/SPT	Asphalt Mixture Performance Tester/Simple Performance Tester
ITS	Indirect tensile strength
CR	Crumb rubber
ISM	Index of stiffness modulus
ITSI	Indirect tensile strength index

References

1. Huang, B.; Shu, X.; Chen, X. Advanced asphalt materials for sustainable pavement construction. *J. Mater. Civ. Eng.* **2020**, *32*, 04020276. [CrossRef]
2. Wang, H.; Li, Z.; You, Z. Performance comparison of porous asphalt mixtures using SBS and rubber-modified bitumen. *Road Mater. Pavement Des.* **2018**, *19*, 111–126.
3. Watson, D.; Qureshi, N.A.; Xie, Z.; Tran, N. *Investigations of Open-Graded Friction Course in Alabama;* National Center for Asphalt Technology (NCAT): Auburn, AL, USA; Alabama Department of Transportation (ALDOT): Montgomery, AL, USA, 2020; NCAT Report 20-07, ALDOT 930-790.
4. Yue, Y.; Abdelsalam, M.; Eisa, M.S. Aggregate Gradation Variation on the Properties of Asphalt Mixtures. *Coatings* **2022**, *12*, 1608. [CrossRef]
5. Xu, G.X.; Kunru, L.; Chunxi, L.; Wang, H.; Leng, Z.; Chen, X. Noise reduction performance and maintenance time of porous asphalt pavement. *Constr. Build. Mater.* **2024**, *452*, 138913. [CrossRef]
6. Lu, X.; Isacsson, U. Modification of road bitumens with thermoplastic polymers. *Polym. Test.* **1998**, *17*, 409–416. [CrossRef]
7. Airey, G.D. Rheological properties of styrene-butadiene-styrene polymer-modified road bitumens. *Fuel* **2003**, *82*, 1709–1719. [CrossRef]

8. González, A.; Poulikakos, L.; Porot, L.; Stempihar, J. Cost-benefit analysis of polymer-modified asphalt in road construction. *Transp. Res. Rec.* **2021**, *2675*, 117–129.
9. Fang, J.; Yu, R.; Liu, X.; Wu, S. Recycling waste tires in asphalt mixtures: A review. *Resour. Conserv. Recycl.* **2019**, *146*, 152–165. [CrossRef]
10. Shen, D.H.; Amirkhanian, S.N.; Tang, B. Influence of surface area and size of crumb rubber on high-temperature properties of rubberized binders. *Constr. Build. Mater.* **2009**, *23*, 304–310. [CrossRef]
11. Ding, H.; Zhao, H.; Huang, B. Performance evaluation of rubber-modified asphalt binders and mixtures: A comprehensive review. *Constr. Build. Mater.* **2020**, *257*, 119507. [CrossRef]
12. Navarro, F.J.; Partal, P.; Martínez-Boza, F.; Gallegos, C. Rheological characteristics of ground tire rubber-modified bitumens. *Chem. Eng. J.* **2004**, *112*, 89–97. [CrossRef]
13. Li, J.; Wang, L.; Liu, P. Industrial production of rubber-modified bitumen and its performance in asphalt mixtures. *J. Clean. Prod.* **2022**, *334*, 130230. [CrossRef]
14. Zhou, X.; Zhang, Z.; Wu, P. Industrial-scale production of rubber-modified asphalt and its application in pavement engineering. *Int. J. Pavement Res. Technol.* **2017**, *10*, 278–287. [CrossRef]
15. Krol, J.; Czajkowski, P.; Radziszewski, P.; Kowalski, K.J.; Sarnowski, M. Viscoelastic properties of polymer and crumb rubber modified bitumen, laboratory and field investigation. In Proceedings of the International Scientific Conference, Road Research and Administration, Car 2013, Bucharest, Romania, 4–5 July 2013.
16. Swieczko-Zurek, B.; Jaskula, P.; Ejsmont, A.J.; Kedzierska, A.; Czajkowski, P. Rolling resistance and tyre/road noise on rubberized asphalt pavement in Poland. Road Mater. *Pavement Des.* **2017**, *18*, 151–167. [CrossRef]
17. *EN 14023*; Bitumen and Bituminous Binders—Specification Framework for Polymer Modified Bitumens. European Committee for Standardization (CEN): Brussels, Belgium, 2010.
18. Pszczola, M.; Jaczewski, M.; Szydlowski, C.; Judycki, J.; Dolzycki, B. Evaluation of low temperature properties of rubberized asphalt mixtures. *Procedia Eng.* **2017**, *172*, 897–904. [CrossRef]
19. *WT-2, 2010*; Technical Guidelines, Asphalt Pavements on State Roads, Asphalt Mixes. General Directorate for National Roads and Motorways: Warsaw, Poland, 2010. (In Polish)
20. *EN 13108-5*; Bituminous Mixtures—Material Specifications—Part 5: Stone Mastic Asphalt. European Committee for Standardization (CEN): Brussels, Belgium, 2016.
21. *EN 13108-7*; Bituminous Mixtures—Material Specifications—Part 7: Porous Asphalt. European Committee for Standardization (CEN): Brussels, Belgium, 2016.
22. *AASHTO R 30-02*; Standard Practice for Mixture Conditioning of Hot-Mix Asphalt (HMA), Standard Specifications for Transportation Materials and Methods of Sampling and Testing, Part 1B: Specifications, 26th Edition. American Association of State Highway and Transportation Officials (AASHTO): Washington, DC, USA, 2006.
23. *EN 1426*; Bitumen and Bituminous Binders—Determination of Needle Penetration. European Committee for Standardization (CEN): Brussels, Belgium, 2015.
24. *EN 1427*; Bitumen and Bituminous Binders—Determination of the Softening Point—Ring and Ball Method. European Committee for Standardization (CEN): Brussels, Belgium, 2016.
25. *EN 12596*; Bitumen and Bituminous Binders—Determination of Dynamic Viscosity by Vacuum Capillary. European Committee for Standardization (CEN): Brussels, Belgium, 2023.
26. *EN 12697-22*; Bituminous Mixtures—Test Methods—Part 22: Wheel Tracking. European Committee for Standardization (CEN): Brussels, Belgium, 2023.
27. *NCHRP 9-29*; NCHRP Report 513: Simple Performance Tester for Superpave Mix Design, First-Article Development and Evaluation. Transportation Research Board, Project 9-29; National Cooperative Highway Research Program: Washington, DC, USA, 2001.
28. *AASHTO TP 62*; Standard Method of Test for Determining Dynamic Modulus of Hot Mix Asphalt (HMA), Standard Specifications for Transportation Materials and Methods of Sampling and Testing. American Association of State Highway and Transportation Officials (AASHTO): Washington, DC, USA, 2007.
29. *EN 12697-26*; Bituminous Mixtures—Test Methods—Part 26: Stiffness. European Committee for Standardization (CEN): Brussels, Belgium, 2023.

materials

MDPI

Article

Research on Crack Resistance of Foamed Rubber Asphalt Cold Recycled Mixtures Based on Semi-Circular Bending Test

Zhen Shen [1], Shikun Wang [1], Zhe Hu [2,*] and Xiaokang Zhao [2]

[1] Overseas Department of CCCC Construction Group Co., Ltd., Beijing 110110, China
[2] School of Highway, Chang'an University, Xi'an 710064, China; zhaoxk@chd.edu.cn
* Correspondence: huzhe@chd.edu.cn

Abstract: Foamed asphalt cold recycled mixtures can provide an effective approach for the reutilization of reclaimed asphalt pavement (RAP), but conventional asphalt foaming technology primarily exploits matrix asphalt as the raw material. To address this issue, this study explores rubberized asphalt with cold recycling technology to develop a foamed rubber asphalt cold recycled mixture (FRCM). The semi-circular bending (SCB) test was employed to investigate its cracking resistance. Load–crack mouth opening displacement (CMOD)–time curves under various temperatures were analyzed, and digital image technique was resorted to monitor crack propagation and growth rates. Fracture toughness, fracture energy, and flexibility index were compared with those of traditional foamed matrix asphalt cold recycled mixture (FMCM). The results show that, under the same test temperature, the FRCM exhibits slower crack propagation; larger peak load; and higher fracture toughness, fracture energy, and flexibility index in comparison with the FMCM. These improvements are more pronounced at low temperatures. For both mixtures, fracture toughness and fracture energy are decreased with increasing the temperature, while the flexibility index shows the opposite trend. The rigid zone accounts for a larger portion of fracture energy at low temperatures. The findings provide technical references for improving the cracking resistance of cold recycled asphalt layers using rubberized asphalt.

Keywords: foamed rubber asphalt; cold recycled mixture; semi-circular bending test; digital image; cracking resistance

1. Introduction

The high-value and resourceful utilization of ground tire rubber and reclaimed asphalt pavement (RAP) has become one of the effective approaches for sustainable development in road transportation infrastructure, contributing to resource conservation and environmental protection [1,2]. Foamed asphalt cold recycling is a green pavement technology that utilizes reclaimed asphalt pavement materials at ambient temperature, incorporating a certain proportion of foamed asphalt, newly added virgin aggregates, cement, and mineral filler [3,4]. After mixing, compaction, and paving, it forms a structural layer that meets the performance requirements of recycled asphalt pavements. This technology has attracted extensive attention due to its advantages in cost-effectiveness, environmental friendliness, and energy conservation with emission reduction [5]. Currently, foamed asphalt cold recycled mixture is mainly applied in asphalt pavement rehabilitation, particularly in low-traffic flexible base courses and intermediate layers, with limited implementation under heavy traffic conditions [6]. As the conventional binder in foamed asphalt cold recycling, matrix asphalt suffers from deficiencies in pavement performance of recycled

https://doi.org/10.3390/ma18122684

mixtures [7]. During service, issues such as raveling and cracking often occur due to insufficient high- and low-temperature stability, poor aging resistance, and inadequate moisture susceptibility, which severely affect the quality of recycled asphalt pavement projects [8,9]. Therefore, enhancing the road performance of foamed asphalt cold recycled mixture has become a bottleneck for large-scale application and promotion. In highway construction, researchers have attempted to convert waste tires into crumb rubber modified asphalt (CRMA) [10]. Relevant studies have shown that rubberized CRMA pavement exhibits excellent low-temperature crack resistance, aging resistance, and moisture stability, effectively reducing traffic noise and delaying pavement cracking and aging issues [11,12]. However, rubberized asphalt mixture requires higher mixing temperatures during production and construction due to its high viscosity, typically ranging from 170 °C to 190 °C, which is approximately 10–20% higher than the mixing temperatures for conventional asphalt mixtures (usually 150–160 °C), leading to massive carbon emissions and pollutant generation, posing health risks to construction workers [13]. If the advantages of foamed asphalt cold recycling and CRMA technology can be effectively integrated, it would not only reduce greenhouse gas emissions and environmental pollution but also improve the pavement performance of recycled pavement structural layers [14]. This approach could contribute significantly to achieving carbon neutrality and carbon peaking goals in the transportation domain, as the production of 1 t of foamed rubberized asphalt cold recycled mixture has been shown to reduce greenhouse gas emissions by approximately 45.13% and energy consumption by 51.38% compared to conventional hot-mix rubberized asphalt mixtures [15].

At present, the basic test methods for assessing the cracking resistance of asphalt mixtures include the semi-circular bending (SCB) test, thermal stress restrained stress test (TSRST), indirect tensile test (IDT), low-temperature bending beam test (BBT), and disc compact tension (DCT) test [16,17]. Based on fracture mechanics theory, the SCB test employs indicators such as fracture energy, fracture toughness, and flexibility index to assess the cracking characteristics of asphalt mixtures [18]. Additionally, the SCB test specimens are highly adaptable, and the internal stress distribution better reflects the actual stress conditions experienced by pavements [19]. The utilization of energy-based indicators for evaluating low-temperature cracking resistance is considered more comprehensive and reasonable, and the test results also show good correlations with in-service pavement performance [20]. Kavussi et al. [21] evaluated the SCB test results of foamed warm mix asphalt mixtures using the flexibility index, fracture energy, and cracking resistance index, revealing that the crack resistance of the mixtures improved with the incorporation of 30–50% RAP. Algraiti et al. [22] and Jeong et al. [23] investigated the influence of rejuvenators on the crack resistance of foamed asphalt cold recycled mixtures through the SCB test, demonstrating that directly adding rejuvenators into RAP materials significantly enhanced the cracking resistance of the recycled mixtures. Sabouri et al. [24] conducted the SCB test to assess the pavement performance of cold in-place recycling materials, and the results indicated that the fracture index value for energy effectively characterized the cracking resistance of the materials. Compared with the BBT, SCB specimens have a larger cross-sectional area and a slower loading rate, enabling a better representation of internal crack propagation characteristics in pavement structures [25]. In contrast to the TSRST, the SCB test offers superior applicability and operational simplicity, as specimens can be prepared using Marshall or rotary compaction methods or extracted directly from in-service pavements [26]. Furthermore, unlike the IDT and DCT tests, the SCB test effectively mitigates localized deformation damage induced by loading strips and allows for twice the number of specimens to be obtained under equivalent testing conditions [27].

Digital image technology provides an efficient and precise analytical tool for studying the fracture behavior of foamed rubber asphalt cold recycled mixture (FRCM) through the SCB test [28,29]. It offers advantages such as non-contact measurement, high measurement accuracy, and automation [30]. By using digital imaging techniques, the entire process of crack appearance, propagation, and final fracture on the surface of SCB specimens under loading can be captured in real time, revealing the damage evolution characteristics of the materials [31]. For example, Yang et al. [25] investigated the fracture process and damage behavior of fiber modified asphalt mixtures using the SCB test and employed the digital image technique to evaluate the interfacial fracture characteristics. The results showed that the three-dimensional distribution of fibers formed a spatial network inside the mixture, effectively delaying crack propagation and enhancing the cracking resistance. Zielinski et al. [20] employed the SCB test to evaluate the relationship between asphalt concrete aging conditions and its fracture performance. Digital image processing techniques were applied to assist laboratory experiments in interpreting the heterogeneity of asphalt mixtures, and digital image correlation was used to verify the accuracy of the test results. Similarly, Ren et al. [32] utilized digital image correlation techniques in the SCB test to study the fracture behavior of asphalt concrete, identifying critical stages of crack appearance and propagation. Hence, digital image technology enables more accurate analysis of crack propagation and the enhancement effects on cracking resistance, supporting the SCB test with more comprehensive results and facilitating the optimized design of asphalt mixtures for improved fracture performance [33]. Although digital image techniques have been applied in fracture testing of various pavement materials, their utilization in the SCB test for analyzing the fracture behavior of FRCM needs to be further explored.

Therefore, the main objective of this study is to evaluate the crack resistance performance of foamed rubber asphalt cold recycled mixtures (FRCM) and clarify the fracture mechanism using the SCB test combined with digital imaging analysis. The mixtures were prepared based on the specifications for RAP in asphalt pavement recycling technology. The fracture behavior of the SCB specimens was analyzed through the relationships among applied load, crack mouth opening displacement (CMOD), and time. Meanwhile, the crack propagation process was recorded using digital imaging equipment, and the crack growth rate was quantitatively evaluated. Furthermore, the fracture toughness, fracture energy, and flexibility index of FRCM were compared with those of FMCM under varying temperature conditions. The outcome contributes to providing technical evidence supporting the application of cold recycled mixture in durable and sustainable pavement structures.

2. Materials and Methods

2.1. Raw Materials

(1) Foamed asphalt

Two types of asphalt binders were utilized in this study: Donghai 70# matrix asphalt and activated rubberized asphalt. The foaming process was carried out using the WLB10S laboratory foaming device and the WLM30 twin-shaft horizontal mixer, both manufactured by Wirtgen (Windhagen, Germany). Based on previous research, the foaming characteristics of different asphalt were determined by adjusting the foaming temperature and water content in order to determine the optimal foaming conditions for rubberized asphalt. The specific procedures and steps followed the methodology described in the study by Zhao et al. [34]. The foaming performance of the asphalt was evaluated using two key indicators: the expansion ratio (the ratio of the maximum volume of foamed asphalt to the volume of unfoamed asphalt) and the half-life (the time required for foamed asphalt to decrease from its maximum volume to 50% of that value). The rubberized asphalt with superior foaming characteristics was selected, exhibiting an optimal foaming water consumption of 4.0%

and an optimal foaming temperature of 165 °C, corresponding to the expansion ratio of 20.2 times and the half-life exceeding 30 s. In comparison, the Donghai 70# matrix asphalt was foamed at 160 °C with a water consumption of 2.9%, resulting in an expansion ratio of 15.5 times and a half-life of 11.8 s. The foaming characteristics of both types of foamed asphalt can meet the specification requirements of JTG/T 5521-2019 in China [35].

(2) RAP

The RAP was sourced from the surface layer of the asphalt pavement during a major rehabilitation project on a highway section in Shaanxi Province, China. A centrifugal extraction method was utilized to recover the aged asphalt from RAP materials, followed by testing three key asphalt properties and the dynamic viscosity at 60 °C. The results indicated that the softening point and dynamic viscosity increased, while penetration and ductility significantly decreased, suggesting that the original adhesive capability of the aged asphalt was essentially lost. Therefore, the RAP was treated as black aggregate in the mixture design. Gradation analysis was conducted on the extracted RAP aggregates. The contents of the aggregates passing through the 2.36 mm sieve size decreased by 20.4% compared to the lower limit, and the 2.36 mm sieve size increased by 3.5% compared to the upper limit. The overall gradation of the RAP materials did not meet the required gradation range for cold recycled mixtures. This indicates that, under long-term traffic loading, most of the coarse aggregates had fractured into medium and fine aggregates. As a result, additional newly added virgin aggregates were required to adjust the gradation of the recycled mixture to meet the standard. Moreover, technical performance tests were carried out on both the coarse and fine aggregates extracted from the RAP, as shown in Table 1.

Table 1. Technical indicators of the RAP.

Materials Testing Programs			Measured Value	Requirements
RAP	Moisture content (%)		1.03	≤3%
	Asphalt content (%)		3.37	>3%
Coarse aggregates	Needle content (%)	Grain size < 9.5 mm	8.1	≤15%
		Grain size > 9.5 mm	3.9	≤20%
	Crushing value (%)		25.1	surface layer ≤ 26% Other layers ≤ 28%
Medium and fine aggregates	Angularity		50%	≥40%

(3) Newly added virgin aggregates and fillers

To meet the gradation design requirements of the foamed asphalt cold recycled mixture, newly added virgin aggregates and fillers were incorporated. Limestone mineral powder was used as the filler to form a denser aggregate skeleton structure of the cold recycled mixture. Additionally, 42.5 MPa Portland cement was added as an active filler to enhance the bonding strength and accelerate the initial curing of the mixture, thereby enhancing the mechanical properties of the cold recycled mixture.

2.2. Design and Preparation of the Cold Recycled Mixture

According to the JTG/T 5521-2019, the gradation of foamed asphalt cold recycled mixtures is categorized into fine-grained, medium-grained, and coarse-grained types. This study adopted the medium-grained gradation design for the foamed asphalt cold recycled mixture. Based on the gradation characteristics of the RAP, newly added virgin aggregates were introduced accordingly. Additionally, mineral filler and cement were incorporated into the mixture. The final gradation (by mass) of the cold recycled mixture was determined to consist of 63% RAP, 6% mineral powder, 1.5% cement, and 29.5% new aggregates. The detailed materials composition and gradation calculations result are shown in Table 2.

Table 2. Materials composition and gradation.

Sieve Size (mm)	Percent Passing (%)					
	63% RAP	6% Mineral Powder	1.5% Cement	29.5% New Aggregates	Synthetic Gradation	Specification Requirements
26.5	100	100	100	100	100	100
19	98.1	100	100	95.2	97.4	85.0–100
16	96.8	100	100	78.2	91.6	-
13.2	94.1	100	100	60.3	84.6	-
9.5	82.8	100	100	48.5	74.0	55.0–80.0
4.75	42.3	100	100	42.6	46.7	35.0–60.0
2.36	19.9	100	100	38.1	31.3	25.0–45.0
1.18	10.0	100	100	35.4	24.2	-
0.6	5.5	100	100	25.1	18.4	-
0.3	2.0	100	100	16.7	13.7	8.0–22.0
0.15	0.9	100	100	8.0	10.4	-
0.075	0.3	89.5	99.2	2.2	7.7	4.0–12.0

The preparation procedure of the foamed asphalt cold recycled mixture (Figure 1a) is as follows: Firstly, the optimum moisture content of the cold recycled mixture without foamed asphalt was determined to be 6.1%, and the optimum mixing water content was 4.9% in accordance with the compaction method in JTG E40-2007 T0131 [36]. Then, a twin-shaft horizontal mixer was used to carry out a two-stage mixing process: Initially RAP, new aggregates, and fillers were mixed with water, followed by the addition of foamed asphalt, and each mixing stage lasted 60 s. After mixing, Marshall specimens (Figure 1b) were prepared using 75 times compaction on each side and cured at a constant temperature of 60 °C and demolded after cooling to room temperature. Six specimens were prepared for each mixture design, with three specimens used for the splitting test at 15 °C and the other three for the splitting test after 24 h of water immersion. Afterwards, the optimal foamed asphalt content was determined according to the peak values of the splitting strength at 15 °C and the tensile strength ratio (TSR) between dry and wet conditions. Ultimately, the optimum foamed asphalt content for both the FMCM and the FRCM was determined to be 3.0%. Under this condition, the dry splitting strengths were 0.48 MPa and 0.72 MPa, respectively, and the optimal dry–wet TSRs were 86.2% and 91.7%. No visible impurities such as oil or biological contaminants were observed during the preparation of the recycled mixtures.

(a) (b)

Figure 1. Foamed asphalt cold recycled mixture: (**a**) samples; (**b**) specimens.

At the optimal asphalt content of 3%, the FMCM only met the specification for the base and subbase layers under non-heavy and above traffic load conditions (splitting strength at 15 °C ≥ 0.40 MPa and TSR ≥ 75%). Notably, the FRCM can meet the surface layer requirements for the heavy and above traffic load levels (splitting strength at 15 °C ≥ 0.60 MPa and TSR ≥ 80%). Compared with FMCM, the splitting strength at 15 °C and TSR of FRCM increased by 50% and 6.4%, respectively. Therefore, the utilization of CRMA as the binder in cold recycled mixtures demonstrates significant performance advantages.

2.3. SCB Test

The preparation of SCB specimens began with forming cylindrical asphalt mixture samples using the rotary compaction method to achieve a size of Φ150 mm × h150 mm, as shown in Figure 2a. Due to the potential differences in compaction density between the top/bottom surfaces and the middle section, and considering that the surface layers are prone to damage, both the top and bottom layers (each 25 mm thickness) were removed using a cutting machine to ensure consistency. The central portion of the specimens were then cut to obtain four semi-circular specimens (Φ150 mm × h50 mm). A 15 mm length and 1.5 mm width notch was introduced perpendicular to the flat edge of each specimen and aligned with the center to facilitate controlled crack propagation during testing. After specimen preparation, the samples were cured for 28 days under standard conditions at 20 ± 5 °C and 40–70% relative humidity. The SCB test was conducted on a MTS universal testing machine, with three replicate specimens per group, to ensure result reliability. The semi-circular bending specimens and its preparation process are shown in Figure 2b,c.

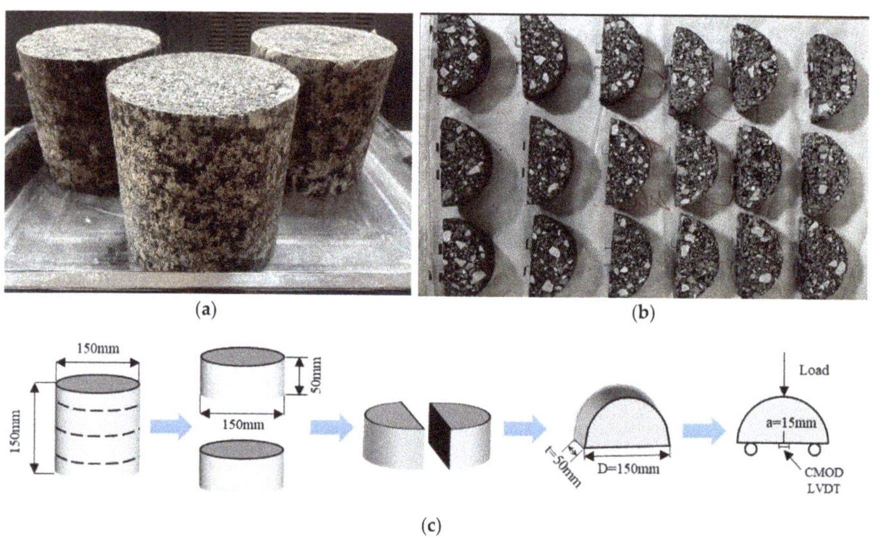

Figure 2. (**a**) Rotary compaction specimen; (**b**) SCB specimen; (**c**) specimen preparation process.

To more clearly highlight the crack propagation procedure of the specimens, the surface of the SCB specimens was sprayed with white paint, as shown in Figure 3. The CMOD of the SCB specimens was measured using a YYU-10/25 electronic extensometer manufactured by Steel Research and Testing Technology Co., Ltd. (Beijing, China), with a measurement range of 10 mm. For accurate attachment, 23/17 staples were adhered to both sides of the notch at the bottom of the SCB specimens using strong adhesive glue. The clips of an extensometer were then secured onto the staples using rubber bands, ensuring stable and reliable displacement measurements during the test.

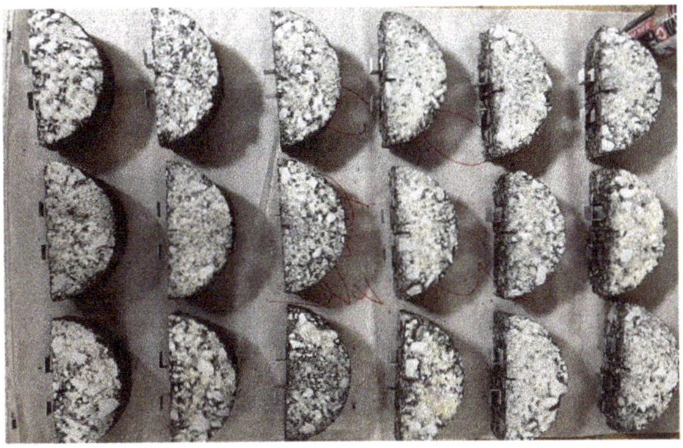

Figure 3. SCB specimens after spray dyeing.

Considering that specimens are prone to brittle fracture and rapid crack propagation at excessively low temperatures, making it difficult to compare the performance of the two types of recycled mixtures, the test temperatures were set at −5 °C, 10 °C, and 25 °C. To ensure thermal equilibrium, the specimens were placed in a MTS temperature-controlled chamber for 6–8 h at each testing temperature prior to the experiment, and the time from the removal of the specimens from the chamber to the start of the SCB test was strictly controlled. The SCB testing procedure is shown in Figure 4. The support points at the bottom of the specimens were aligned to ensure the specimen was centered both laterally and longitudinally, with the distance set at 0.8 diameter (120 mm). Loading was applied using a MTS universal testing machine, with the loading head vertically aligned with the central notch of the specimens. This alignment ensured that the loading direction coincided with the notch of the specimen. The loading rate was set to 2 mm/min, as a slower loading speed can promote crack propagation and damage evolution. In addition, a high-resolution camera was employed to record the loading process, with the camera axis positioned perpendicular to the specimen surface, providing a clear visualization of the crack propagation and fracture characteristics.

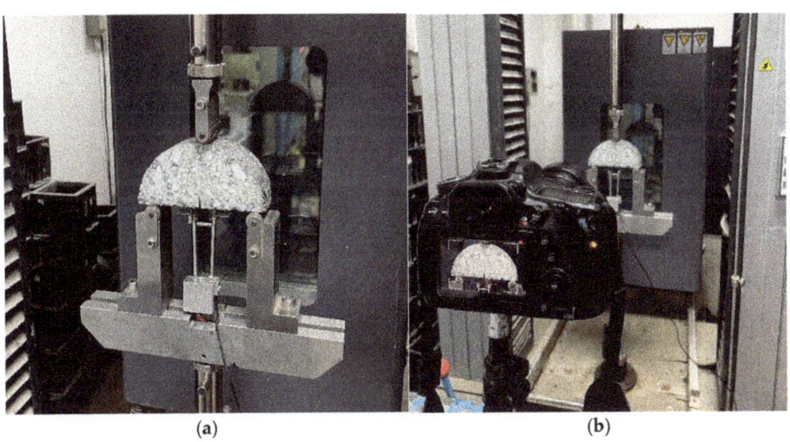

Figure 4. SCB test process: (**a**) extensometer installation and specimen placement; (**b**) camera takes pictures to record the cracking process.

3. Results and Discussion

3.1. Analysis of the Fracture Process and Characterization of the Mixtures Based on the SCB Test

3.1.1. Load–CMOD–Time Curve

This study plotted the load–time and CMOD–time curves for FMCM and FRCM at different temperatures, as shown in Figure 5. It can be observed that, for both mixtures, the load–time relationship exhibited a typical trend of increasing to a peak followed by a decline, indicating a complete loading and failure process. Under the 25 °C test conditions, both types of asphalt mixtures mainly exhibit viscoelastic behavior, with the load reaching its peak after more than 40 s. After reaching the peak load, visible microcracks gradually began to appear at the bottom notch of the SCB specimen, and the CMOD started to change. After the peak load, the microcracks progressively extend to form cracks, the strength of the mixture begins to degrade, and the load drops to zero. Eventually, cracks propagated through the entire specimen, leading to ductile failure. Under the 10 °C condition, the peak load for both mixtures is greater than at 25 °C, and the peak load occurs earlier, between 30 and 40 s. This can be attributed to the increase in stiffness of the mixtures at lower temperatures. When the test temperature was further dropped to −5 °C, a significant increase in the peak load was observed, and the peak load occurred earlier around 20–30 s. After the peak load, the load decreased rapidly, indicating that the specimen experienced brittle fracture behavior. The load–displacement curve is not completely smooth. This is because microcracks generated inside the specimen accumulate along the edges of the aggregates, leading to macroscopic microcracks and abrupt load fluctuations.

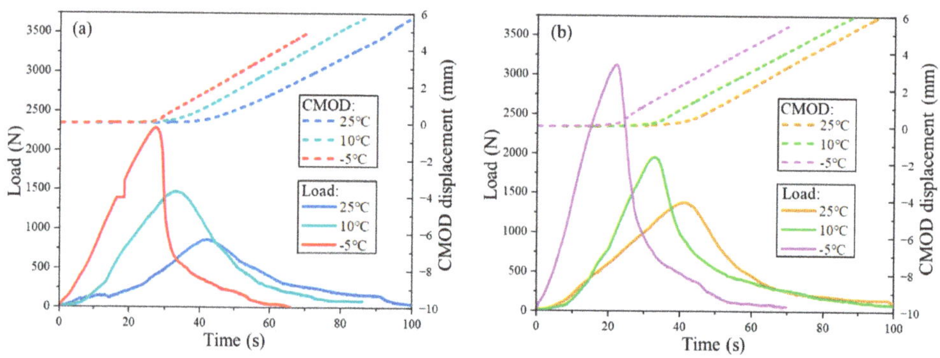

Figure 5. Load and CMOD–time curves at different temperatures: (**a**) FMCM; (**b**) FRCM.

The comparison of the fracture processes of the two types of asphalt mixtures reveals that, under the same temperature, the critical failure load of the FRCM is significantly higher than that of the FMCM. This can be attributed to the stronger adhesion property of rubberized asphalt, which results in a more robust bond with the aggregates. An analysis of the CMOD–time curves for the two mixtures shows that, before the peak load, the CMOD remains essentially unchanged and close to 0, indicating no significant deformation in the SCB specimen. After the peak load, the CMOD begins to increase, showing a linear growth trend. Moreover, the CMOD evolution curves of the two mixtures under different temperatures are nearly parallel, indicating a similar rate of crack opening. This suggests that the resistance mechanisms of the FRCM and FMCM in the crack propagation process are similar, but the FRCM demonstrates a greater crack resistance, with its peak load being approximately 36.4% and 33.3% higher than that of the FMCM at −5 °C and 10 °C.

3.1.2. Crack Fracture Process and Propagation Rate

According to different stress states and fracture characteristics, crack propagation can be classified into three basic modes: crack opening mode (I), crack sliding mode (II), and tearing mode (III). The fracture states of the SCB specimen after loading are shown in Figure 6. Under the applied load, the tensile stress is generated at the specimen bottom, and cracks initially form at the tip of the notch. The crack then propagates upwards along the original notch direction, eventually penetrating through the entire specimen. Therefore, the fracture mode is identified as the crack opening mode (I). The end of the crack was defined as the moment when the primary crack extended through the entire ligament of the SCB specimen, accompanied by a significant drop in load (approaching zero) on the load–CMOD–time curve. This was further confirmed visually through digital imaging. Figure 6 also shows a representative cracked specimen with the end of the crack indicated.

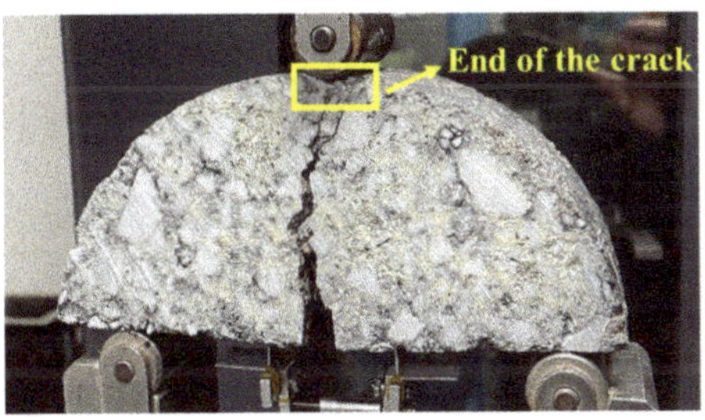

Figure 6. Fracture state of the SCB specimen.

Furthermore, a Canon digital device with high resolution and a sufficient frame rate was utilized during the experiment to continuously record the entire cracking development process of the SCB specimen. The cracking process of the FRCM specimen under the −5 °C test condition at different time points was recorded, as shown in Figure 7.

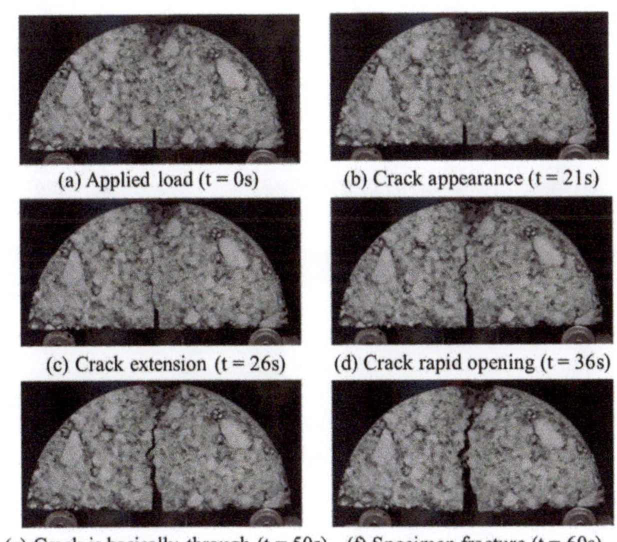

Figure 7. Foamed rubber asphalt cold recycled mixture specimen crack propagation fracture process at −5 °C.

Figure 7 demonstrates that, during the loading phase, no visible changes occurred on the surface of the specimen. After 21 s of loading, internal microcracks had already formed, and fine cracks began to appear on the surface. As the loading continued, the load reached its maximum value and began to decay, and the crack gradually expanded. The crack propagation mainly occurred at the bonding interface between the aggregates and the asphalt, and the cracks tended to circumvent the surfaces of the larger aggregates. As the loading head continued to descend, the crack mouth opening displacement continued to increase. Around 60 s, the crack had propagated through the entire specimen, signifying that the SCB specimen had fractured.

In order to better understand the crack propagation procedure of the SCB specimens under a load, the crack propagation distance per unit time was calculated. The crack propagation velocity was determined using the following formula, as shown in Equation (1):

$$v = \frac{\Delta s}{\Delta t} \tag{1}$$

where v is the crack propagation speed (mm/s); Δs is the crack propagation path length (mm); Δt is the crack propagation time (s).

Potplayer V1.7 software was resorted to capture real-time video frames during the specimen loading process. The extracted images were then imported into ImageJ V1.8 software for path calibration and crack length measurement, as shown in Figure 8. The distance between the two bottom support points was selected as the calibration basis, and the scale was set to 0.8 in diameter (120 mm). The Segmentedline tool in ImageJ was employed to trace the crack propagation location, and the measured path length was output as the crack propagation length at the time of the SCB specimen fracture. By displaying the real-time video, the time from crack appearance to specimen fracture was recorded, and the crack propagation rate was calculated.

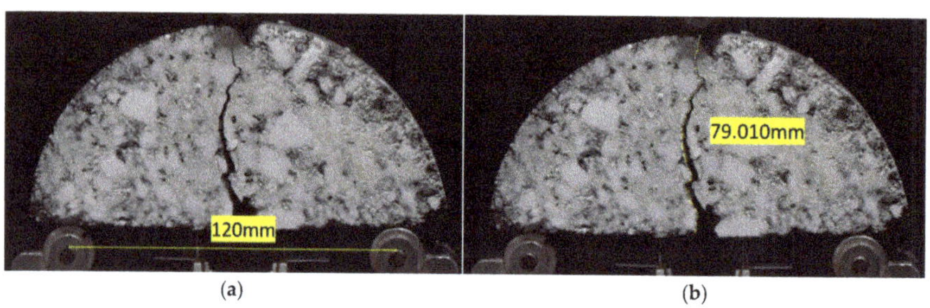

Figure 8. Crack propagation path calibration and measurement for the SCB specimen: (**a**) size calibration; (**b**) path measurement.

Figure 9 shows the crack propagation rate test results for various specimens at different temperatures. The results indicate that, when the test temperature decreases, the crack propagation rate of both types of cold recycled mixtures gradually increases. This is attributed to the fact that asphalt mixtures tend to exhibit brittle behavior at lower temperatures, thus leading to faster crack propagation. However, the crack propagation rate of FRCM shows almost no difference between 10 °C and 25 °C, which may be attributed to the fact that the fracture process is dominated by material toughness in this temperature range. In addition, rubber particles in the FRCM play a significant toughening role and exhibit strong temperature adaptability. Within the temperature range of 10 °C to 25 °C, the rubber particles can maintain good elasticity, continuously absorbing and dissipating part of the energy during crack propagation, thereby stabilizing the crack propagation rate.

Moreover, the crack propagation rate of the FMCM was higher than that of the FRCM for all three test temperatures, indicating that the FRCM exhibits better toughness and crack resistance. This can be attributed to the reality that rubber particles can reduce the stress concentration at the crack tip, resulting in higher resistance to crack propagation in the recycled mixture.

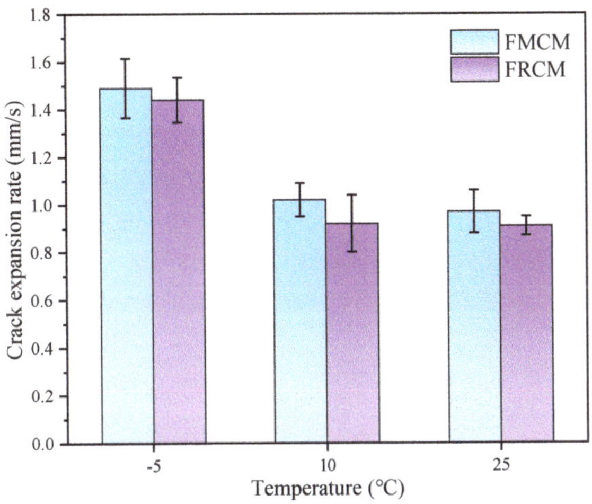

Figure 9. Crack propagation rate of the SCB specimens.

3.2. Analysis of the Crack Resistance of the Mixtures Based on the SCB Test

3.2.1. Fracture Toughness

The fracture toughness K_{IC} corresponds to the stress intensity factor K_I at the peak load *Pc*, and it can be utilized to characterize the stress distribution at the crack tip during the cracking process of the asphalt mixtures specimens. When the stress intensity factor of the material K_I reaches the fracture toughness K_{IC}, visible cracks begin to form. As the loading continues, these cracks rapidly propagate. Therefore, a larger K_{IC} value indicates a higher stress intensity at the maximum load, meaning the asphalt mixture absorbs more energy during the fracture procedure. This results in better resistance to crack propagation. The equations for calculating the fracture toughness K_{IC} can be seen as follows:

$$K_{IC} = Y_{I(0.8)}\sigma_0\sqrt{\pi a} \quad (2)$$

$$Y_{I(0.8)} = 4.782 + 1.219\left(\frac{a}{r}\right) + 0.063e^{7.045\left(\frac{a}{r}\right)} \quad (3)$$

$$\sigma_0 = \frac{P_c}{2rt} \quad (4)$$

where K_{IC} is the fracture toughness (MPa·m$^{0.5}$); $Y_{I(0.8)}$ is the standard stress intensity factor, dimensionless; P_c is the peak load (mN); σ_0 is the stresses corresponding to the peak load (kN); *a* is the depth of the cut at 0.015 m; *r* is the radius of the SCB specimen at 0.075 m; *t* is the thickness of the SCB specimen at 0.05 m.

The fracture toughness results for FMCM and FRCM at three different temperatures are shown in Figure 10. The results indicate that the fracture toughness of both cold recycled mixtures is highly influenced by temperature. The fracture toughness of both mixtures increases when the temperature decreases. This trend is especially evident between 10 °C and −5 °C, where the fracture toughness rises significantly by more than 50% for FRCM and FMCM, with increases of 60.7% and 57.1%, respectively. This can be ascribed to the reality that a decrease in temperature has a stiffening effect on asphalt, leading to a higher load-bearing capacity and higher fracture toughness of the asphalt mixture.

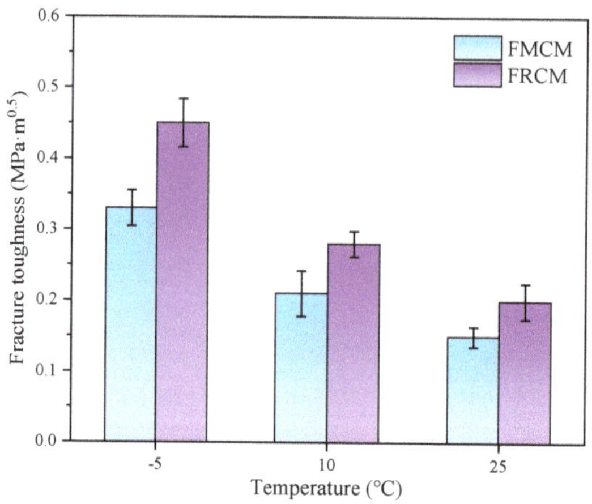

Figure 10. Fracture toughness of the specimens at different temperatures.

Moreover, the fracture toughness of both cold recycled mixtures at −5 °C is 2.25 times and 2.2 times higher than at 25 °C, respectively, indicating that more energy is required for fracture propagation at lower temperatures. Notably, at the same temperature, the fracture toughness of the FRCM is consistently higher than that of the FMCM, with this advantage being particularly prominent at low temperatures. Hence, the results of the fracture toughness suggest that the FRCM has better low-temperature crack resistance than the FMCM. However, it should be noted that fracture toughness depends on the magnitude of the peak load and does not account for the displacement factor during the fracture of SCB specimens. As a result, it may not fully reflect the comprehensive performance differences between various asphalt mixtures.

3.2.2. Fracture Energy

Fracture energy G_f can also be utilized to assess the cracking resistance of asphalt mixtures. It is the ratio of the fracture work to the crack propagation area, and the larger G_f indicates the better cracking resistance of the mixture. The specific calculation is as follows:

$$G_f = \frac{W_f}{A_{lig}} \tag{5}$$

where G_f is the fracture energy (J/m^2); W_f is the fracture work (J), $W_f = \int pd(u)$; A_{lig} is the crack propagation area (m^2), $A_{lig} = (r-a)t$.

In addition, the fracture work refers to the work done by the external force on the SCB specimens of the cold recycled mixture. It can be calculated as the integral of the load–vertical displacement curve along the x-axis, from the onset of deformation to the point where the residual load drops to 100 N. The specific calculation formula is as follows:

$$W_f = \int_0^{u_0} P_1(u)du + \int_{u_0}^{u_{final}} P_2(u)du \tag{6}$$

where $P_1(u)$ is the pre-peak fitting curve; $P_2(u)$ is the post-peak fitting curve; u_0 is the displacement of the peak load (mm); u_{final} is the displacement of the 0.1 kN residual load (mm).

As shown in Figure 11, the peak load serves as a boundary, dividing the curve into two regions in the load–vertical displacement curve: the rigid and the ductile regions. In this study, the load–displacement curve was fitted using software Origin 2024, and the area enclosed by the fitted curve and the x-axis was integrated to calculate the fracture work of the cold recycled mixtures under different temperatures. Then, the fracture energy was obtained by calculating the ligament area based on the specimen thickness and ligament length. Three parallel tests were performed under each test condition, and the average fracture energy value was taken to ensure the reliability of the experimental data.

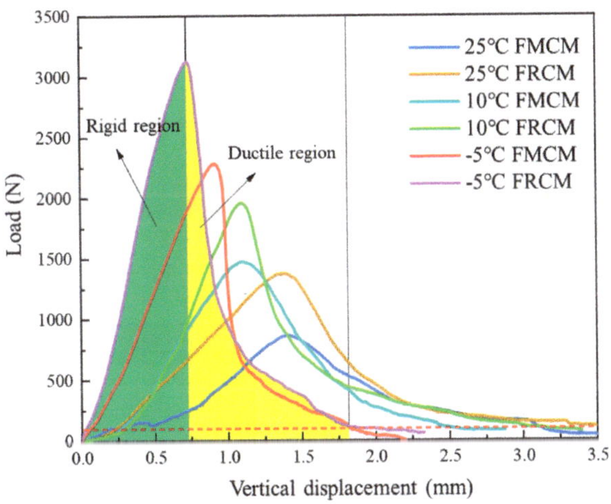

Figure 11. Load–vertical displacement curve.

The fracture energy of the FMCM and FRCM at different temperatures is shown in Figure 12. Compared to the matrix asphalt, rubberized asphalt can effectively enhance the cracking resistance of the cold recycled mixture. As the temperature increases, the fracture energy in the rigid region of both mixtures decreases. Compared with the FMCM, the fracture energy in the rigid region of the FRCM is increased by 13.8%, 25.3%, and 67.8%, respectively, at the test temperatures of −5 °C, 10 °C, and 25 °C, and the fracture energy in the ductile region is increased by 75.3%, 28.6%, and 41.5%, respectively.

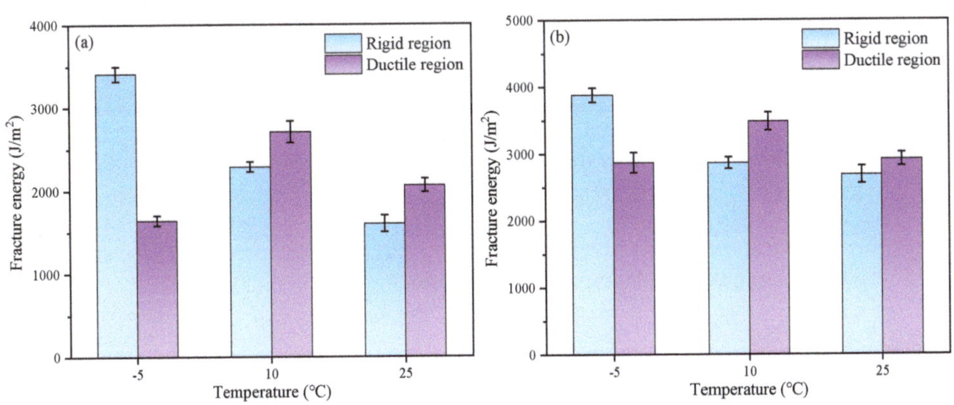

Figure 12. Fracture energy of the specimens at different temperatures: (**a**) FMCM; (**b**) FRCM.

Additionally, the fracture energy in the rigid and ductile regions was calculated separately, and the proportion of each region is shown in Figure 13. The findings indicate that the trend of fracture energy in the rigid region of both asphalt mixtures is consistent with the overall fracture energy trend. Under the −5 °C low temperature, the SCB specimen exhibits a high peak load and is prone to brittle failure. Therefore, after reaching the peak load, the load decreases rapidly, resulting in a larger proportion of fracture energy in the rigid region before the peak load. When the test temperature increases, the stiffness of the asphalt mixture decreases, and its ductility increases, causing the proportion of fracture energy in the rigid region to decrease. At low temperatures, the matrix asphalt exhibits a higher degree of brittleness compared to rubberized asphalt, which also contributes to a significantly smaller proportion of fracture energy in the rigid region for the FRCM compared to the FMCM. However, due to the higher strength of the FRCM, its fracture energy in the rigid region is still larger than that of the FMCM. Under the 10 °C and 25 °C test conditions, there is little difference in the stiffness of the FRCM and FMCM, and the proportions of fracture energy in the rigid region for both cold recycled mixtures are quite similar.

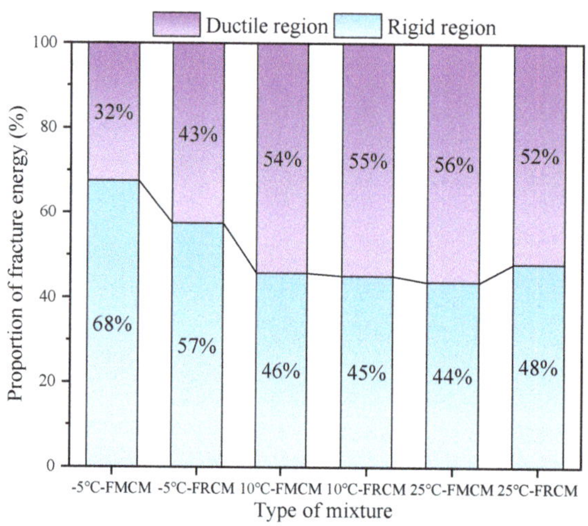

Figure 13. Proportions of the fracture energy in the rigid and flexible zones.

3.2.3. Flexibility Index

The fracture energy is insufficient to evaluate asphalt mixtures with the same fracture energy and different load–displacement characteristics [37]. The flexibility index can be calculated by the ratio of fracture energy to the slope at the inflection point of the load–displacement curve after the peak load, which can effectively evaluate the crack propagation process. The slope at the inflection point after the peak load can effectively characterize the unloading rate during the crack appearance to full propagation and can also reflect the extent of brittle fracturing during the unloading process. The specific calculation formula is as follows:

$$FI = \frac{G_f}{|m|} * A \quad (7)$$

where FI is the flexibility, dimensionless; m is the value of slope at the inflection point of the load displacement curve after the peak load; A is the calibration factor coefficient, and the default value is 0.01 (kN/mm^2).

The flexibility index results of the specimens at different test temperatures are shown in Table 3. The results indicate that the flexibility index increases significantly when the temperature increases. A higher value of the flexibility index represents slower crack propagation and stronger resistance to cracking. At low temperatures, the specimens exhibit more brittle behavior, so the crack propagation rates of the FMCM and FRCM are nearly identical, and the values of the flexibility index show little difference. When the test temperature rises from −5 °C to 10 °C, and then from 10 °C to 25 °C, the growth rates (slopes) of the flexibility index for the FMCM are 0.582 and 1.369, respectively. For the FRCM, the growth rates (slopes) are 1.678 and 2.201, respectively. The average growth rate of the flexibility index for the FRCM is approximately twice that of the FMCM as the temperature increases from −5 °C to 25 °C. Therefore, it can be concluded that, compared to the FMCM, the FRCM exhibits slower crack propagation and better resistance to cracking damage. This is consistent with the results obtained from the fracture toughness and fracture energy indicators. Furthermore, as the temperature increases, the improvement in mixture toughness and the enhancement of the crack resistance performance become more pronounced in the FRCM.

Table 3. Flexibility index of the SCB specimens at different temperatures.

Type of Mixture	Flexibility Index		
	−5 °C	10 °C	25 °C
FMCM	1.26	9.99	30.52
FRCM	4.09	29.26	62.28

4. Conclusions

This study combined two environmentally friendly asphalt cold recycling and rubberized asphalt technologies to prepare the FRCM and systematically evaluated its cracking resistance compared to traditional FMCM through the SCB test with digital imaging equipment under different temperature conditions. The following conclusions were drawn from the study:

(1) The FRCM can satisfy the technical design requirements of the surface layers under medium and heavy traffic, while the FMCM only meets the requirements for the base layers. At all three test temperatures, the peak load of the FRCM is 33.3–36.4% higher than that of the FMCM, indicating an improved load-bearing capacity.

(2) Compared to the FMCM, the FRCM exhibits a slower crack propagation rate and greater flexibility index, indicating better resistance to crack growth. As the temperature increases from −5 °C to 25 °C, the average growth rate of the flexibility index for the FRCM is approximately twice that of the FMCM.

(3) The fracture toughness of both the FRCM and FMCM increases as the temperature decreases, with a particular rise observed between −5 °C and 10 °C. In this range, the fracture toughness of the FRCM and FMCM increases by 60.7% and 57.1%, respectively, indicating enhanced crack resistance at lower temperatures.

(4) The fracture energy of the FRCM is higher than that of the FMCM under all three test temperatures, indicating better crack resistance. The proportion of fracture energy in the rigid region exceeds 50% at low temperatures.

This study utilizes digital image techniques to assist the SCB test, which is currently limited to performance validation in practical engineering and macroscopic-scale evaluations. Future research may establish a multiscale evaluation for the FRCM by integrating the macroscopic mechanical performance with microscopic structural characterization and further investigate its high-temperature rutting resistance, moisture susceptibility, and durability. In addition, developing a carbon footprint assessment framework and conducting a cost–benefit analysis would further promote the advancement of this environmentally friendly and sustainable pavement technology.

Author Contributions: Investigation, S.W. and Z.H.; Methodology, Z.S. and X.Z.; Conceptualization, Z.H. and X.Z., Validation, Z.S. and S.W.; Writing—original draft preparation, Z.S. and Z.H.; Writing—review and editing, X.Z. and S.W.; Funding acquisition, X.Z. All authors have read and agreed to the published version of the manuscript.

Funding: This work was supported by the Shaanxi Natural Science Foundation (Grant No. 2024JC-YBMS-373). The authors gratefully acknowledge their financial support.

Institutional Review Board Statement: Not applicable.

Informed Consent Statement: Not applicable.

Data Availability Statement: The original contributions presented in this study are included in the article. Further inquiries can be directed to the corresponding author.

Conflicts of Interest: Authors Zhen Shen and Shikun Wang were employed by the company Overseas Department of CCCC Construction Group Co., Ltd. The remaining authors declare that the research was conducted in the absence of any commercial or financial relationships that could be construed as a potential conflict of interest.

References

1. Bilema, M.; Aman, M.Y.; Hassan, N.A.; Al-Saffar, Z.; Mashaan, N.S.; Memon, Z.A.; Milad, A.; Yusoff, N.I.M. Effects of Waste Frying Oil and Crumb Rubber on the Characteristics of a Reclaimed Asphalt Pavement Binder. *Materials* **2021**, *14*, 3482. [CrossRef] [PubMed]
2. Farina, A.; Kutay, M.E.; Anctil, A. Environmental Assessment of Asphalt Mixtures Modified with Polymer Coated Rubber from Scrap Tires. *J. Clean. Prod.* **2023**, *418*, 138090. [CrossRef]
3. Buczynski, P.; Sramek, J.; Mazurek, G. The Influence of Recycled Materials on Cold Mix with Foamed Bitumen Properties. *Materials* **2023**, *16*, 1208. [CrossRef] [PubMed]
4. Hu, Z.; Wei, Z.; Zhao, X.; Zhang, M.; Zhang, J.; Pei, J.; Wang, Q.; Lyu, L. Foamed Waste Oil-Activated Rubberized Asphalt Binder: A Sustainable Recycling Approach for Improving Foaming Effect and Performance. *Constr. Build. Mater.* **2024**, *423*, 135889. [CrossRef]
5. Liu, Q.; Pei, J.; Li, R.; Zhao, X.; Meng, Y.; He, H.; Song, J. Influence of Fatty Acid Amide Foaming Agents on the Foaming Effect and Rheological Properties of Foamed Asphalt. *Constr. Build. Mater.* **2025**, *458*, 139721. [CrossRef]
6. Chen, Z.; Yi, J.; Zhao, H.; Luan, H.; Xu, M.; Zhang, L.; Feng, D. Strength Development and Deterioration Mechanisms of Foamed Asphalt Cold Recycled Mixture Based on MD Simulation. *Constr. Build. Mater.* **2021**, *269*, 121324. [CrossRef]
7. Mugume, R.B. Investigation of Foamed Bitumen Mixes Using Reclaimed Asphalt Pavement Materials for Cold Recycling Technology. *Int. J. Pavement Res. Technol.* **2022**, *15*, 98–110. [CrossRef]
8. He, Y.; Hu, Q.; Zhao, X.; Zhang, J.; Lyu, L.; Li, Y.; Aldagari, S.; Fini, E.H. Impacts of Ultraviolet Irradiation on Asphalt Binder: A Comprehensive Review. *Int. J. Pavement Eng.* **2025**, *26*, 2472839. [CrossRef]
9. Hu, Z.; Gu, R.; Zhang, M.; Zhang, J.; Lyu, L.; Wang, X.; Li, J.; Pei, J. Synergistic Mechanism of Calcium Sulphate Whisker and Polyurethane on Asphalt Binder: Experimental and Computational Investigation. *Road Mater. Pavement Des.* **2025**. [CrossRef]
10. Thives, L.P.; Pais, J.C.; Pereira, P.A.A.; Palha, C.A.O.F.; Triches, G. Contribution of Asphalt Rubber Mixtures to Sustainable Pavements by Reducing Pavement Thickness. *Materials* **2022**, *15*, 8592. [CrossRef]
11. Li, Y.; Zhao, C.; Li, R.; Zhang, H.; He, Y.; Pei, J.; Lyu, L. Dry-Process Reusing the Waste Tire Rubber and Plastic in Asphalt: Modification Mechanism and Mechanical Properties. *Constr. Build. Mater.* **2025**, *458*, 139759. [CrossRef]
12. Lin, Y.; Cai, J.; Ma, W.; Bi, Y.; Ji, X.; Gu, F.; Yu, H. Optimal Design and Validation of Double-Layer Porous Asphalt Structure Based on Thin-Layer Process. *Constr. Build. Mater.* **2025**, *467*, 140362. [CrossRef]
13. Huang, G.; Chen, Z.; Bi, Y.; Wang, S.; Zhang, J.; Falchetto, A.C. A Comprehensive Evaluation of Crumb Rubber Modified Asphalt-Aggregate Interfacial Behavior and the Effect of Water: Molecular Dynamics Simulation and Experimental Investigation. *Fuel* **2024**, *377*, 132824. [CrossRef]
14. Wen, Y.; Wang, Z.; Dong, Y.; Pei, J.; Zhang, J. Preparation and Performance Evaluation of Foamed Terminal Blend Rubberized Asphalt Binders. *J. Build. Mater.* **2023**, *26*, 1129–1136. [CrossRef]

15. Hu, Z.; Zhang, J.; Yu, X.; Zhao, X.; Lyu, L.; Wang, Q.; Zhang, M.; Huang, G. Laboratory Evaluation of Cold Recycled Mixture with Foamed Waste Oil-Activated Rubberized Asphalt. *Transport. Res. Part D-Transport. Environ.* **2024**, *135*, 104395. [CrossRef]
16. Salamat-Talab, M.; Zeinolabedin-Beygi, A.; Soltani, F.; Akhavan-Safar, A.; Carbas, R.J.C.; da Silva, L.F.M. Tensile Strength and Mode I Fracture Toughness of Polymer Concretes Enhanced with Glass Fibers and Metal Chips. *Materials* **2024**, *17*, 2094. [CrossRef]
17. Zhou, B.; Pei, J.; Zhang, J.; Guo, F.; Wen, Y.; Luo, P. Comparison of Fracture Test Methods for Evaluating the Crack Resistance of Asphalt Mixture. *Arab. J. Sci. Eng.* **2020**, *45*, 8745–8758. [CrossRef]
18. Zarei, M.; Kordani, A.A.; Zahedi, M. Evaluation of Fracture Behavior of Modified Warm Mix Asphalt (WMA) under Modes I and II at Low and Intermediate Temperatures. *Theor. Appl. Fract. Mech.* **2021**, *114*, 103015. [CrossRef]
19. Xue, B.; Pei, J.; Zhou, B.; Zhang, J.; Li, R.; Guo, F. Using Random Heterogeneous DEM Model to Simulate the SCB Fracture Behavior of Asphalt Concrete. *Constr. Build. Mater.* **2020**, *236*, 117580. [CrossRef]
20. Zielinski, P.; Klimczak, M.; Tekieli, M.; Strzepek, M. An Evaluation of the Fracture Properties of Asphalt Concrete Mixes Using the Semi-Circular Bending Method and Digital Image Correlation. *Materials* **2025**, *18*, 967. [CrossRef]
21. Kavussi, A.; Motevalizadeh, S.M. Fracture and Mechanical Properties of Water-Based Foam Warm Mix Asphalt Containing Reclaimed Asphalt Pavement. *Constr. Build. Mater.* **2021**, *269*, 121332. [CrossRef]
22. Algraiti, W.; Kavussi, A. The Role of Rejuvenating Agents in Cold Recycled Foam Asphalt Mixes. *Case Stud. Constr. Mater.* **2023**, *19*, e02515. [CrossRef]
23. Jeong, M.M.; Kim, S.; Shen, J. Laboratory Evaluation of Foamed Asphalt Mixtures with 100% RAP and Rejuvenator. *Aust. J. Civ. Eng.* **2021**, *19*, 46–57. [CrossRef]
24. Sabouri, M.; Wegman, D.E. Performance Evaluation of Cold In-Place Recycling Materials through a Simple Semi-Circular Bending Test. *Road Mater. Pavement Des.* **2023**, *24*, 1158–1172. [CrossRef]
25. Yang, Y.; He, Y.; Fu, R.; Zhao, X.; Shang, H.; Ma, C. Understanding Toughening Mechanisms and Damage Behavior in Hybrid-Fiber-Modified Mixtures Using Digital Imaging. *Buildings* **2024**, *14*, 2562. [CrossRef]
26. Pszczola, M.; Szydlowski, C. Influence of Bitumen Type and Asphalt Mixture Composition on Low-Temperature Strength Properties According to Various Test Methods. *Materials* **2018**, *11*, 2118. [CrossRef]
27. Falchetto, A.C.; Moon, K.H.; Wang, D.; Riccardi, C.; Wistuba, M.P. Comparison of Low-Temperature Fracture and Strength Properties of Asphalt Mixture Obtained from IDT and SCB under Different Testing Configurations. *Road Mater. Pavement Des.* **2018**, *19*, 591–604. [CrossRef]
28. Cunha, F.G.; Santos, T.G.; Xavier, J. In Situ Monitoring of Additive Manufacturing Using Digital Image Correlation: A Review. *Materials* **2021**, *14*, 1511. [CrossRef]
29. Chomicz-Kowalska, A. A Study of Adhesion in Foamed WMA Binder-Aggregate Systems Using Boiling Water Stripping Tests. *Materials* **2021**, *14*, 6248. [CrossRef]
30. Liu, T.; Li, Y.; Chen, Z.; Zhang, J.; Lyu, L.; Pei, J. Damage Evolution in Asphalt Mixtures Based on In-Situ CT Scanning. *Constr. Build. Mater.* **2024**, *438*, 137266. [CrossRef]
31. Yuan, F.; Cheng, L.; Shao, X.; Dong, Z.; Zhang, L.; Wu, G.; He, X. Full-Field Measurement and Fracture and Fatigue Characterizations of Asphalt Concrete Based on the SCB Test and Stereo-DIC. *Eng. Fract. Mech.* **2020**, *235*, 107127. [CrossRef]
32. Ren, H.; Qian, Z.; Huang, W.; Bo, W.; Chen, T.; Cao, H. Evaluation of Fracture Behavior in Asphalt Concrete through the Combination of Semi-Circular Bending Test and Digital Image Correlation Technology. *Constr. Build. Mater.* **2024**, *451*, 138854. [CrossRef]
33. Pei, Z.; Lou, K.; Kong, H.; Wu, B.; Wu, X.; Xiao, P.; Qi, Y. Effects of Fiber Diameter on Crack Resistance of Asphalt Mixtures Reinforced by Basalt Fibers Based on Digital Image Correlation Technology. *Materials* **2021**, *14*, 7426. [CrossRef] [PubMed]
34. Zhao, X.; Ma, C.; Chen, L.; Zhang, J.; Hu, Z. Foaming characteristics and optimum foaming conditions of waste rubber asphalt. *J. Chang. Univ. Sci. Technol. (Nat. Sci.)* **2024**, *21*, 95–106. [CrossRef]
35. *JTG/T 5521-2019*; Technical Specifications for Highway Asphalt Pavement Recycling. Ministry of Transport of the People's Republic of China: Beijing, China, 2019.
36. *JTG E40-2007*; Test Methods of Soils for Highway Engineering. Ministry of Transport of the People's Republic of China: Beijing, China, 2007.
37. Mehdinejad, S.; Fazaeli, H.; Moniri, A.; Dabiri, A.S. Comparison of Two Criteria of Stress Intensity Factor and Fracture Energy to Investigate the Behavior of Asphalt Mixtures under Combined Tensile-Shear Loading Modes-A Statistical Approach. *Constr. Build. Mater.* **2021**, *290*, 123230. [CrossRef]

materials

Article

A Study on the Road Performance of the Self-Healing Microcapsule for Asphalt Pavement

Pei Li [1], Rongyi Ji [2], Chenlong Zhang [1], Jinghan Xu [2], Mulian Zheng [2,*] and Xinghan Song [2]

1 Guangdong Road and Bridge Construction Development Co., Ltd., Guangzhou 510635, China; lipeijiaoer@163.com (P.L.); zhchl1987@126.com (C.Z.)
2 Key Laboratory for Special Area Expressway Engineering of Ministry of Education, Chang'an University, Xi'an 710064, China; jirongyi@chd.edu.cn (R.J.); 2021021051@chd.edu.cn (J.X.); songxinghan@chd.edu.cn (X.S.)
* Correspondence: zhengml@chd.edu.cn; Tel.: +86-17753355660

Abstract

Asphalt pavement cracking is an important factor affecting its service life. Under certain conditions, the self-healing behavior of asphalt itself can repair pavement cracks. However, the self-healing ability of asphalt itself is limited. In order to strengthen the self-healing ability of asphalt, the microcapsule wrapped with a repair agent is pre-mixed into the asphalt mixture. When the crack occurs and spreads to the surface of the microcapsule, the microcapsule ruptures and the healing agent flows out to realize the self-healing of the crack. Current microcapsules are mostly prepared with healing agents and bio-oil as core materials, and their high-temperature resistance to rutting is poor. While the epoxy resin contains a three-membered cyclic ether, it can undergo ring-opening polymerization to bond and repair the asphalt matrix. In addition, research on microcapsules mainly focuses on the self-healing properties of microcapsule-modified asphalt. In fact, before adding microcapsules to asphalt to improve its self-healing performance, it is necessary to ensure that the asphalt has a good road performance. On this basis, the self-healing performance of asphalt is improved, thereby extending the service life of asphalt pavement. Therefore, two-component epoxy self-healing microcapsules (E-mic and G-mic) were first prepared in this paper. Then, a temperature scanning test, rheological test of bending beams, and linear amplitude scanning test were, respectively, conducted for the microcapsule/asphalt to evaluate its road performance, including the high-temperature performance, low-temperature crack resistance, and fatigue performance. Finally, the self-healing performance of microcapsules/asphalt was tested. The results showed that the self-developed epoxy self-healing microcapsules were well encapsulated and presented as spherical micron-sized particles. The average particle size of the E-mic was approximately 23.582 μm, while the average particle size of the G-mic was approximately 22.440 μm, exhibiting a good normal distribution. In addition, they can remain intact and unbroken under high-temperature conditions. The results of road performance tests indicated that the microcapsule/asphalt mixture exhibits an excellent high-temperature resistance to permanent deformation, low-temperature crack resistance, and fatigue resistance. The self-healing test demonstrated that the microcapsule/asphalt exhibited an excellent self-healing performance. When the microcapsule content was 4%, the self-healing rate reached its optimal level of 67.8%, which was 149.2% higher than that of the base asphalt.

Keywords: asphalt pavement; micro-cracks; microcapsule; road performance; self-healing property; epoxy resin; two-component

1. Introduction

Asphalt pavement can repair micro-cracks and local damage by its self-healing ability under normal temperatures and in service conditions. While, the self-healing ability is limited and the speed is very slow [1,2], many researchers proposed using additives to enhance the self-healing ability. Diao et al. demonstrated that the fatigue self-healing efficiency of asphalt concrete could be improved by using slaked lime as a filler [3–5]. Zhang et al. found that molecular and nano-level polymers could enhance the self-repairing property of asphalt mixtures [6]. Jin et al. studied the impact of adding self-healing microcapsules on asphalt performance. Their research results showed that incorporating microcapsules into asphalt can effectively enhance its self-healing properties. As the number of microcapsules increases, the self-healing index of microcapsule-containing asphalt first increases and then decreases [7]. Ding et al. established a macrostructure model for asphalt pavement and a multiscale model of the representative volume element (RVE) incorporating self-healing microcapsules using the finite difference method. They systematically analyzed the effects of the microcapsule content, capsule wall modulus, and capsule wall thickness on the mechanical properties of different asphalt pavements. The results indicate that the microcapsule content, microcapsule wall material modulus, and wall membrane thickness have a significant impact on the road performance and self-healing ability of asphalt pavement [8]. After developing self-healing microcapsules, Wang et al. analyzed the impact of microcapsules on the self-healing and fatigue life of asphalt and found that when the dosage was 2%, the improvement effect was better [9]. Zhang et al. synthesized MMF microcapsules through in situ polymerization and analyzed the improvement effects of adding the microcapsules on the temperature sensitivity, self healing ability, and fatigue life of asphalt [10].

While microcapsule-modified asphalt significantly improves the self-healing efficiency and fatigue life of asphalt, its incorporation may exert complex impacts on other pavement performances of asphalt [11–14]. Wen et al. prepared microcapsules with different core–shell ratios by coating soybean oil and melamine–urea–formaldehyde (MUF). The microcapsules were characterized by a particle size analysis, a morphology observation, and other methods to determine the optimal core–shell ratio. The results showed that the microcapsules had a good particle state when the core–shell ratio was 1:2. The viscosity of asphalt increases after the incorporation of microcapsules. Upon the rupture of the microcapsules, the release of the soybean oil alters the composition ratio of the asphalt, leading to a decrease in the high-temperature performance. However, the remaining microcapsule shells continue to enhance the deformation resistance of asphalt [15]. Song et al. prepared urea–formaldehyde–epoxy resin (UFE) microcapsules using polyurea formaldehyde as the wall material and epoxy resin as the core material and concluded that UFE improved both the high-temperature deformation resistance and self-healing ability of SBS-modified asphalt [16]. Wang and his team discovered that modifying the microcapsule shell with 1 wt% graphene improved the high-temperature rutting resistance of asphalt, indicating that microcapsules containing graphene are beneficial [17]. However, the incorporation of microcapsules has a negative impact on the low-temperature performance of asphalt. When the content of microcapsules is too high, the ductility of the modified asphalt will be unqualified [10,18]. In order to improve the low-temperature crack resistance of the asphalt binder, researchers usually modify microcapsules. Wen et al. adopted carbon nanotube-modified microcapsules [19], and Ren et al. adopted graphene-modified microcapsules [20]. It was found that the low-temperature performance of the microcapsule-modified asphalt was improved.

At present, there are mainly three kinds of materials used as the core materials of microcapsules for asphalt, including asphalt-based materials, biological oil (Bio-oil), and adhesive materials [21–23]. Asphalt-based materials mainly included an asphalt rejuvenator and a light oil [24,25]. Asphalt-based materials have a poor high-temperature performance. When they are applied to pavement, their resistance to permanent deformation becomes worse, and the fatigue life decreases. Biological oil belongs to environmentally friendly materials, which are conducive to energy conservation and environmental protection to reduce emissions. However, its viscosity is low, and the fluidity is high, resulting in easy overflows under high temperatures [26]. Adhesive materials mainly include epoxy resin and waterborne epoxy resin. Adhesive materials repair micro-cracks through the curing and bonding effect, which can be regarded as the repair of cracks in the real sense. Nowadays, the molecular structure of the ordinary epoxy resin used contains a significant amount of unsaturated bonds, which are easy to degrade under ultraviolet radiation and have a poor anti-aging performance [27]. The current research on microcapsule asphalt mainly focuses on self-healing properties. The incorporation of microcapsules into asphalt can be considered as a modification of the material itself. The prerequisite for enhancing the self-healing performance of asphalt is to have an ideal road performance, at least not sacrificing other performances.

In this paper, two-component microcapsules, E-mic and G-mic, were first prepared using modified epoxy resin E and curing agent G, respectively, as the core materials of the microcapsules, both of which exhibit excellent weather resistance. Secondly, the microcapsules prepared were added to the asphalt to prepare the microcapsules/asphalt material. Meanwhile, temperature scanning, a low-temperature bending creep test, and a linear amplitude scanning test were conducted to evaluate the road performance of the microcapsule/asphalt. Finally, the self-healing property test was performed to evaluate the self-healing property of the microcapsule/asphalt. The research process is illustrated in Figure 1.

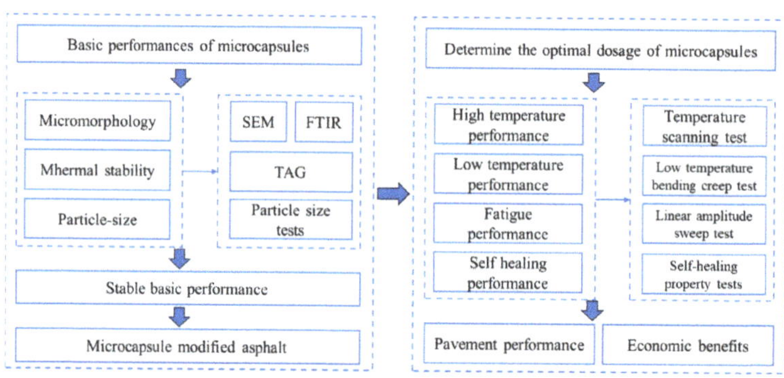

Figure 1. The research process of this paper.

2. Test Materials and Methods

2.1. Test Materials

2.1.1. Asphalt

The asphalt adopted the Korean SK70 base asphalt. The technical indicators are shown in Table 1.

Table 1. The indicators of SK 70.

Indicators	Unit	Measured Value	Specified Value
Penetration (25 °C, 100 g, 5 s)	0.1 mm	66.7	60–80
Ductility (5 cm/min)	cm	>100	≥100
Softening point	°C	46.7	≥42/43
Viscosity (135 °C)	mPa s	475	-

2.1.2. Microcapsules

(1) Core Material

Ordinary epoxy resin materials contain a large number of unsaturated linkages, which will be cracked under ultraviolet light. Based on this, epoxy resin E with better aging resistance and weather resistance was selected as the epoxy component of epoxy self-healing microcapsules. Epoxy resin E is a colorless and transparent low-viscosity liquid. It is obtained by modifying ordinary bisphenol A epoxy resin under high-temperature and high-pressure conditions. The molecule of E is a six-membered ring without double bonds and unsaturated bonds and has an extremely low chlorine content. Its structural formula is shown in Figure 2. E can be cured under low-temperature, medium-temperature, and UV light conditions, and its cured product has strong aging resistance and weather resistance. Its main technical indicators are listed in Table 2.

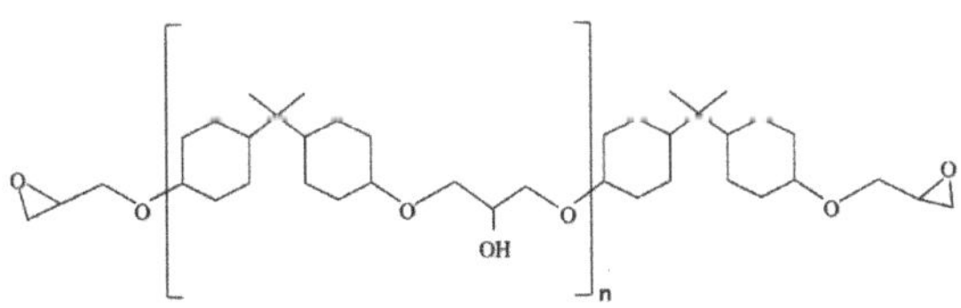

Figure 2. The chemical structure of E. Reprinted with permission from Ref. [23]. Copyright 2020 Elsevier.

Table 2. Main technical indicators of epoxy resin E.

Viscosity/mPa·s/25 °C	Epoxy Equivalent/g·eq^{-1}	Color/Max, G	Performance Overview
2300	215	1	Obtained by hydrogenation modification of bisphenol A type, colorless and transparent, weather-resistant, and yellowing-resistant.

A medium-to-low temperature curing agent G, based on a mixed modified amine system, was selected as the curing agent. Its main components include triethylenetetramine (TETA) and diethylenetriamine (DETA), etc. It can initiate curing reactions at temperatures above 0 °C and effectively repair pavement damage within the typical service temperature range of pavements. The curing process proceeds in two steps: first, the epoxy groups of the epoxy resin open and undergo grafting reactions with the curing agent to form hydroxyl groups; second, the grafted products further crosslink with unreacted epoxy groups, eventually forming a three-dimensional network structure. The specific reaction process is shown in Figure 3.

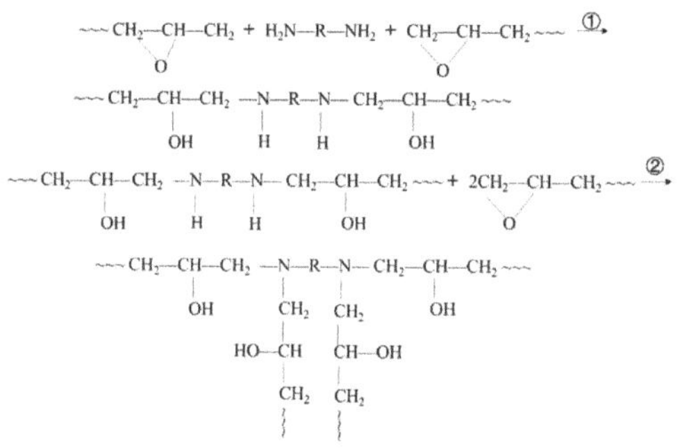

Figure 3. Curing reaction of epoxy resin E and curing agent G.

(2) Shell Material

Common non-biodegradable polymer shell materials for microcapsules include melamine formaldehyde resin (HMMM), melamine modified urea–formaldehyde resin (MUF), etc. Among them, HMMM, despite its excellent chemical stability, tends to react prematurely with the core material epoxy resin E, which may cause microcapsule failure. In contrast, MUF not only has lower cost but also exhibits good thermal stability and mechanical strength. In this study, MUF was selected as the shell material for microcapsules. Table 3 lists the raw materials of MUF. The pre-polymer can produce condensation polymerization reaction with an acidic catalyst and at a certain temperature, thus forming the polymer MUF. This reaction is shown in Figure 4.

Table 3. Raw materials for MUF preparation.

Material	Molecular Formula	Specification
Melamine (M)	$C_3N_3(NH_2)_3$	AR
Urea (U)	CH_4N_2O	AR
Formaldehyde (F)	CH_2O	AR

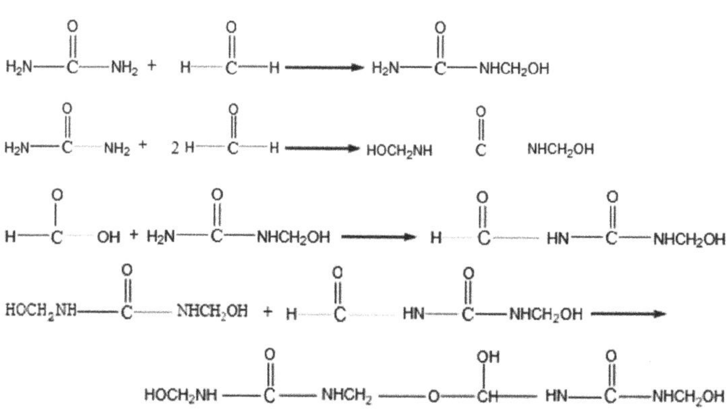

Figure 4. Synthetic reaction of MUF.

(3) Preparation of Microcapsules

To prepare microcapsules, the first step is to prepare the shell material (MUF) of the microcapsules. Subsequently, the core material (E/G) was emulsified. Afterwards, the MUF was added into the core material emulsion drop by drop. And then, the curing reaction was conducted for 3 h after adding the catalyst, facilitating the curing reaction.

Finally, the microcapsules were obtained after washing with distilled water and filtering. The process is shown in Figure 5.

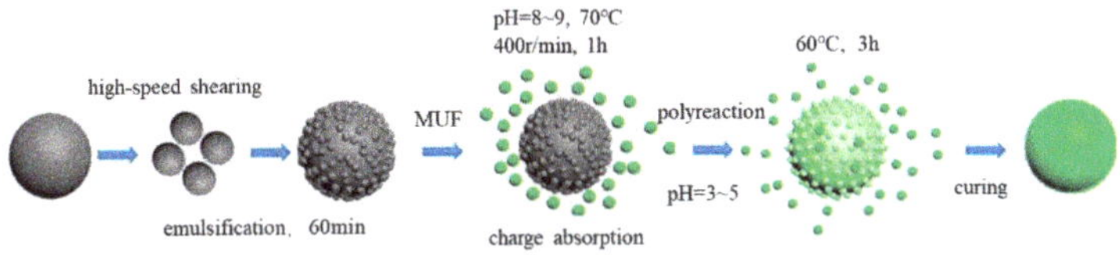

Figure 5. Microencapsulation process. Reprinted with permission from Ref. [23]. Copyright 2020 Elsevier.

2.1.3. Microcapsule/Asphalt

Asphalt was heated to 160 °C ± 5 °C, and microcapsules E-mic and G-mic were sequentially added to the asphalt. The microcapsule/asphalt was stirred with a blender at 500 r/min for 30 min to prevent uneven microcapsule particle agglomeration and dispersion. The samples were prepared at the dosages of 1%, 2%, 3%, and 4%, respectively, and the dosage was the mass ratio of E-mic to asphalt (the mass ratio of E-mic to G-mic was 2:1). Therefore, the four groups of microcapsule/asphalt samples were, respectively, denoted by 1% E-mic, 2% E-mic, 3% E-mic, and 4% E-mic in this study.

2.2. Test Methods

2.2.1. Performance Tests of the Microcapsule

The microstructure and thermal stability of the microcapsule will directly affect its performance in asphalt and asphalt mixture. In this study, micromorphology, particle size distribution, and thermal stability tests were carried out, respectively.

(1) Micromorphology and Microstructure Tests

Micromorphology and microstructure of the microcapsules were observed by SEM. The SEM was produced by Carl Zeiss AG, Jena, Germany (model MERLIN Compact). Before conducting observations, it was necessary to immobilize the microcapsule powder and endow it with conductivity (by spraying gold on its surface). Subsequently, an electron beam scanning observation was performed with a scanning voltage of 5 kV and a current of 10 μA, thereby obtaining the surface micromorphology and microstructure of the microcapsules.

(2) Particle Size Tests

The instrument used for particle size distribution testing was sourced from the UK and was modeled as the Nastersizer 2000 (Malvern Panalytical, Malvern, UK). The dispersing agent was ethanol with a refractive index of 1.330, particle absorptivity of 0.100, and particle refractive index of 1.587. The scattering model was Mie, the analysis model was universal, and the analysis sensitivity was set to enhancement.

(3) Thermogravimetric Analysis (TGA) Tests

Microcapsules used for modifying asphalt must possess good high-temperature stability to ensure that most of the capsules remain intact after high-temperature mixing and construction. Therefore, TGA is used to test the state of microcapsules at high temperatures. The Synchronous thermal analyzer adopted is produced in America, and its model is Discovery SDT 650 (TA Instruments, New Castle, DE, USA).

About 10 mg microcapsule samples were weighed and loaded into alumina crucible, and the heating process was carried out in an N_2 flow of 40 mL. A stepwise isothermal heating mode was adopted in order to capture the mass change more accurately. The heating rate was 10 °C·min^{-1}, and the heating range was 40~600 °C.

(4) Fourier Transform Infrared Spectroscopy (FTIR) Tests

Infrared absorption spectroscopy images can reveal the functional group information of a substance. By utilizing FTIR to test the changes in functional groups before and after asphalt modification, one can infer the chemical reactions that occurred during the modification process [28–30]. The test was conducted using a Tensor II infrared spectrometer produced in Germany. The resolution was set to 4 cm^{-1}, and the test wavenumber ranged from 400 to 4000 cm^{-1}.

2.2.2. Temperature Scanning (TS) Test

Temperature scanning test was used to analyze the high-temperature rheological properties of microcapsule/asphalt to evaluate its resistance to permanent deformation at high temperature. The samples were tested using a SmartPave 102 (Anton Paar, Graz, Austria). A parallel plate of 8 mm was used with a 2 mm plate spacing. The tests adopted strain control mode, and the strain was set at 12%. The angular frequency was set at 10 rad/s. The test temperature was set at 52 °C, 58 °C, 64 °C, and 70 °C.

2.2.3. Low-Temperature Bending Creep Stiffness Test

The low-temperature bending creep stiffness test was used to evaluate the cracking resistance of microcapsules/asphalt at low temperature. The low-temperature bending beam rheometer (BBR) is produced in the United States, model TE-BBR-F. The samples were prepared as asphalt beams with a size of 127 mm × 12.7 mm × 6.35 mm. Three parallel samples were prepared for each condition. The beam should be preloaded to ensure its close contact with the bracket. Afterwards, a constant load of 980 mN ± 25 mN was applied to the beam for 240 s at a given temperature. Meanwhile, the central deformation of the beam was measured by a displacement sensor. The creep stiffness modulus (S) and the creep velocity (m) of the beam at 8.0 s, 15.0 s, 30 s, 120 s, and 240 s were recorded to analyze the cracking resistance of microcapsules/asphalt at low temperature. The test temperature was set to −12 °C, −18 °C, and −24 °C.

2.2.4. Linear Amplitude Sweep (LAS) Test

The linear amplitude sweep (LAS) tests were used to analyze the effect of microcapsules on the fatigue resistance of matrix asphalt. The same DSR as in the previous TS test was employed for this test. PAV aging was performed on the prepared microcapsule/asphalt before sample preparation. The samples prepared were 8 mm in diameter and 2 mm in spacing, and the test temperature was set at 25 °C. The LAS test consists of two steps according to AASHTO TP 101-14 [31]. The first step is to analyze the rheological properties of samples by frequency scanning. The frequency scanning adopts the controlled strain mode, and the strain is set to 0.1%. The test frequency is set to 0.2, 0.4, 0.6, 0.8, 1.0, 2.0, 4.0, 6.0, 8.0, 10, 20, and 30 Hz. The complex modulus and phase angle at each frequency are recorded during the tests. The second step is to perform an amplitude scanning test. The test frequency (ω) is set at 10 Hz, and the sinusoidal load amplitude is linearly increased from 0.1% to 30%. During the test, the time interval of data recording is 1 s. The amplitude scanning test is loaded for 3100 cycles with a duration of 310 s.

After the frequency scanning test, the storage modulus $G'(\omega)$ was obtained. $\log \omega$ was used as the horizontal axis and $\log G'(\omega)$ as the vertical axis, and linear fitting was performed by Equation (1). Then, parameter α can be calculated by Equation (2).

$$\log G'(\omega) = m(\log \omega) + b \tag{1}$$

$$\alpha = 1/m \tag{2}$$

The accumulated damage of the sample after the amplitude scanning test can be calculated by Equation (3).

$$D(t) \cong \sum_{i=1}^{N} \left[\pi\gamma_0^2(C_{i-1} - C_i)\right]^{\alpha/1+\alpha} (t_i - t_{i-1})^{1/1+\alpha} \tag{3}$$

where $|G*|$ denotes complex shear modulus (MPa); γ_0 denotes the strain at a given point (%); t denotes test time (s); $C(t) = |G^*|(t)/|G^*|_{initial}$ is the $|G*|$ at time t divided by the initial undamaged $|G*|$ (MPa). The value of the $|G*|$ is the value of the second point, because the modulus of the first point after changing the condition from the static state was different from the undamaged modulus at the target loading frequency. At time t, $C(t)$ and $D(t)$ are recorded, assuming that $C = 1$ when $D(0) = 0$. The relationship between $|G*| \sin\delta$ and $D(t)$ is shown in Equation (4).

$$C(t) = C_0 - C_1(D)^{C_2} \tag{4}$$

where $C_0 = 1$. C_1 and C_2 are the coefficients derived from curve fitting using Equation (5).

$$\log(C_0 - C(t)) = \log(C_1) + C_2 \cdot \log(D(t)) \tag{5}$$

D_f denotes the $D(t)$ corresponding to fatigue damage, and can be calculated by Equation (6).

$$D_f = \left(\frac{C_0 - C_{at\ Peak\ Stress}}{C_1}\right)^{1/C_2} \tag{6}$$

Parameters A and B of the fatigue damage model of asphalt can be calculated by Equation (7).

$$A = \frac{f\left(D_f\right)^k}{k(\pi C_1 C_2)^\alpha} \tag{7}$$

where f denotes loading frequency (10 Hz); $k = 1 + (1 - C_2)\alpha$; and $B = 2\alpha$.

Fatigue performance parameter N_f of asphalt can be expressed in Equation (8).

$$N_f = A(\gamma_{\max})^{-B} \tag{8}$$

where $\gamma_{\max}$ denotes the maximum strain of a given pavement structure (%).

2.2.5. Self-Healing Property Tests

"Fatigue–Rest–Fatigue" test based on DSR was designed to evaluate the self-healing property of microcapsules/asphalt in this study. A parallel plate of 8 mm was used with a 2 mm plate spacing. The loading frequency was set at 10 Hz. The stress was set to 3%, and the temperature was set at 25 °C. Relevant studies have shown that the complex shear modulus (G*) drops to 70% of the initial modulus (G_0^*), which can better represent the self-healing performance of asphalt materials [32]. Therefore, in this experiment, the loading was stopped when G* dropped to 70% of G_0^*, completing the first "fatigue" test. Subsequently, the specimens were kept at 25 °C for 1 h, that is, the "rest" period to repair the fatigue damage. Afterwards, tests continued until G* dropped again to 70% of G_0^*. It was the second "fatigue" test. "Fatigue–Rest–Fatigue" test is illustrated in Figure 6.

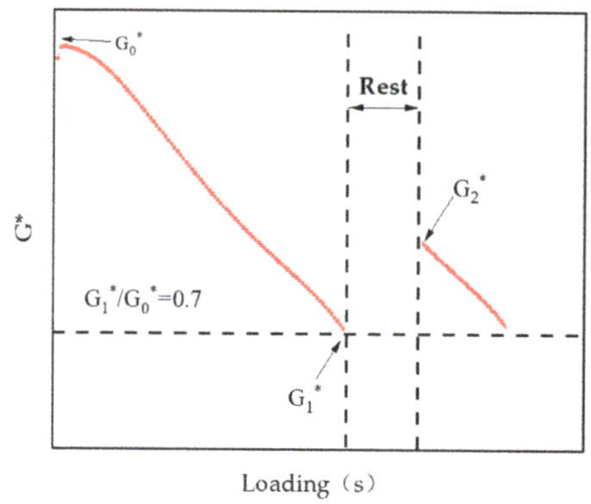

Figure 6. A schematic diagram of the "Fatigue–Rest–Fatigue" test.

The self-healing rate of microcapsule/asphalt samples can be calculated by Equation (9).

$$SH = \frac{G_2^* - G_1^*}{G_0^* - G_1^*} \times 100\% \quad (9)$$

where SH denotes the self-healing rate of microcapsule/asphalt; G_0^* denotes the initial complex shear modulus; G_1^* denotes the complex shear modulus before the repair period; G_2^* denotes the complex shear modulus after the repair period.

3. Results and Discussions

3.1. Basic Performances of Microcapsules

The microcapsules prepared in the laboratory are in the form of a dry powder, where the E-mic is pure white and the G-mic is milky white, as shown in Figure 7. Both types of microcapsule powders exhibit a uniform particle dispersion without obvious agglomeration, which is closely related to the in situ polymerization process adopted. During preparation, the core materials are first dispersed into micron-sized droplets by emulsifiers. The droplets are then modified to carry positive charges via the addition of the 0.5 wt% cationic surfactant. Subsequently, the shell precursor undergoes in situ polymerization on the surface of positively charged core droplets due to electrostatic adsorption, forming microcapsules with a core–shell structure. Finally, dry sieving yields powder products with an excellent dispersibility.

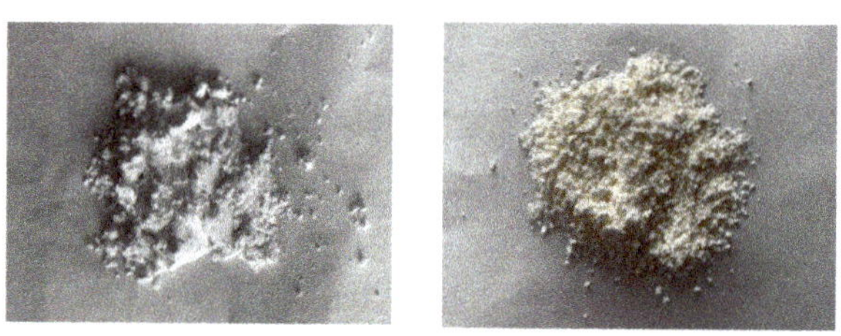

(**a**) E-mic (**b**) G-mic

Figure 7. Microcapsule samples.

3.1.1. Micromorphology and Microstructure Tests

The microstructure and morphology of the microcapsules were observed via SEM. Microscopic images are presented in Figures 8 and 9.

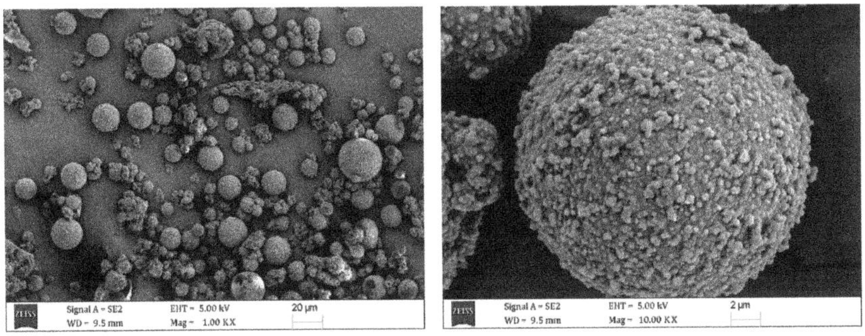

Figure 8. SEM images of E-mic.

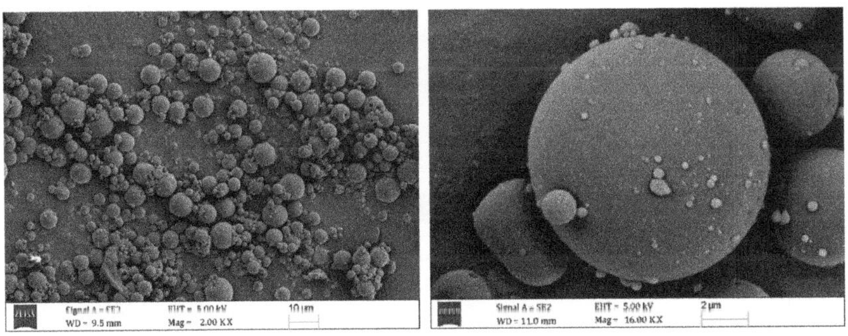

Figure 9. SEM images of G-mic.

As can be seen from Figure 8, the shape of the E-mic was close to spherical, with MUF particles adhering to its surface and a compact encapsulation. The relatively rough surface of the E-mic could improve its contact state with the matrix asphalt, making the modified asphalt system more stable. When asphalt sustains minor damage, epoxy resin E could flow out smoothly to repair the damage.

Figure 9 shows the micromorphology and microstructure of the G-mic. As depicted in Figure 9, the surface of the G-mic was slightly smoother than that of the E-mic. Similarly to the E-mic, the G-mic exhibited a good surface condition, with tightly packed and evenly distributed particles. This was conducive to the dispersion of microcapsules in asphalt, thereby enhancing the performance stability of the microcapsule-modified asphalt.

3.1.2. Particle Size Analysis

The prepared microcapsules were powdery solids at the macroscopic level. The particle size analysis results of microcapsules are shown in Figure 10.

As shown in Figure 10a, the particle size distribution curve of the E-mic sample shows an approximately normal distribution, and the size is concentrated in a region ranging from 5 to 110 μm. Similarly to the E-mic, the particle size distribution curve of the G-mic sample also shows an approximately normal distribution, and the size is concentrated in a region ranging from 5 to 100 μm. The samples within each size range are analyzed, and the analysis results are obtained in Table 4.

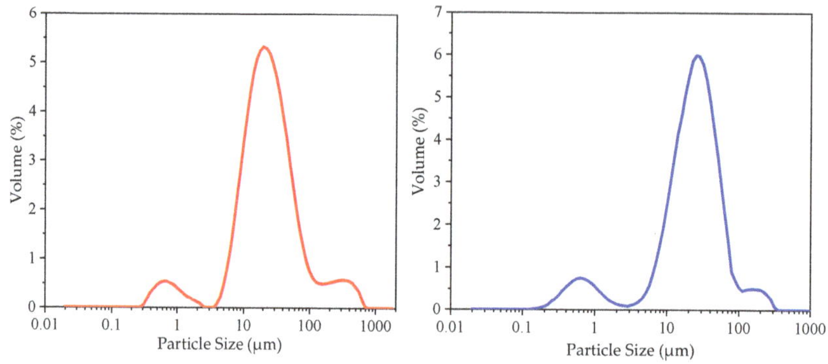

(**a**) E-mic (**b**) G-mic

Figure 10. The particle size distribution of the microcapsules prepared.

Table 4. The results of the particle size analysis.

Samples	D(10) (μm)	D(50) (μm)	D(90) (μm)	Span	Consistency	Shading Degree
E-mic	7.096	23.582	87.440	1.113	0.343	10.93%
G-mic	4.477	22.440	63.246	1.157	0.355	12.90%

As listed in Table 3, D(x) denotes the particles with a size smaller than D(x), which accounts for x% of the total number of particles, where x = 10, 50, and 90. Meanwhile, D50 also denotes the mean particle size. The consistency indicates the uniformity of the particle size distribution. The mean particle size of the E-mic sample was 23.582 μm, and the particle size of more than 90% was below 87.440 μm; the mean particle size of the G-mic sample was 22.440 μm, and the particle size of more than 90% was below 63.246 μm. From the particle size distribution test results, it can be seen that the particle size distributions of the two types of microcapsules are uniform, exhibiting good homogeneity.

3.1.3. Microcapsule Thermal Stability Analysis

The thermal stability of microcapsules was tested by a thermogravimetric analyzer. The TGA of the E-mic and G-mic is shown in Figure 11.

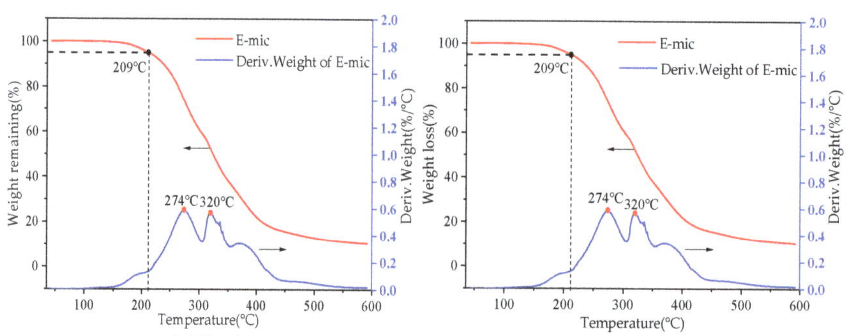

(**a**) E-mic (**b**) G-mic

Figure 11. TGA curve for E-mic and G-mic.

The TG curve represents the weight loss in the test, and the DTG curve is the derivative of the TG curve, that is, the change rate of the weight loss with the temperature. As shown in Figure 11a, the temperature at which the residual weight of the E-mic reaches 95%

is 209 °C. This indicates that MUF begins to thermally decompose at 209 °C. When the temperature reaches 274 °C, the decomposition rate of MUF accelerates, indicating that at this temperature, the outer shell of MUF is essentially ineffective. As the temperature continues to rise to approximately 320 °C, a second peak appears on the DTG curve. Epoxy resin E also enters a rapid decomposition stage. As shown in Figure 11b, for the G-mic, the initial decomposition temperature is 223 °C, the first peak of the DTG is 275 °C, and the second peak is 382 °C.

Therefore, the E-mic and G-mic encapsulated by MUF have a high-heat resistance and can survive at high temperatures during asphalt construction without decomposition.

3.1.4. Analysis of Interaction Mechanism Between Microcapsules and Asphalt

The infrared spectrum of a substance is a reflection of its molecular structure, and the absorption peak in the spectrum corresponds to the vibration form of each group in the molecule. Figure 12 shows the corresponding infrared spectra of the G-mic after it is added to the base asphalt. The characteristic absorption peaks of C-H and C=O are near 2920 cm^{-1} and 2850 cm^{-1}. The stretching vibration characteristic peak of C=O (in ester, aldehyde, ketone, acid) is near 1733 cm^{-1}; the stretching vibration peak of C=C is near 1680 cm^{-1}; the characteristic peak of the vibration absorption of the benzene ring is near the wavenumber of 1600 cm^{-1} is. The characteristic peaks of the asymmetric variable angle vibration and the symmetrical variable angle vibration of CH_3 are near 1454 cm^{-1} and 1376 cm^{-1}, respectively. The characteristic peak of the C-N stretching vibration is near 1239 cm^{-1} in the G-mic. The anti-symmetric stretching characteristic peak of the ether bond (C-O-C) in the G-mic is near 1182 cm^{-1}. Based on the above analysis, this indicates that the G-mic is physically mixed with the asphalt, but there is no chemical reaction between them.

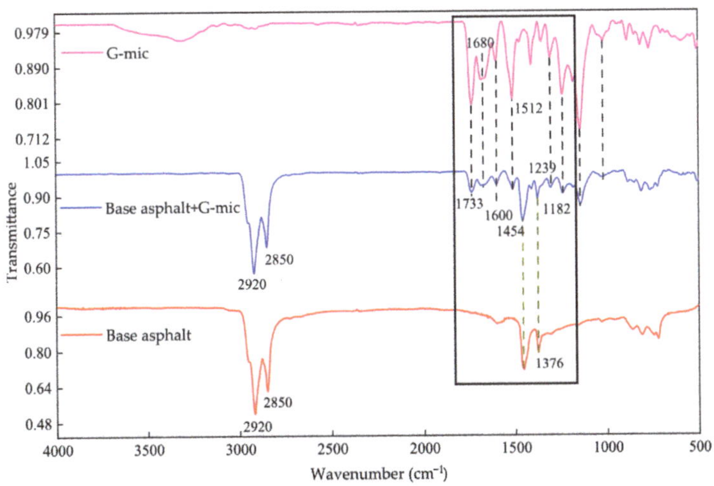

Figure 12. The infrared spectrum of the base asphalt and the base asphalt + G-mic.

Figure 13 shows the corresponding infrared spectra of the E-mic added to the base asphalt. Compared with the G-mic absorption peak, the N-H bending vibration absorption peak of the E-mic at 1500 is weakened, and the epoxy functional group is added at 916. This is related to the different functional groups of E-mic and G-mic molecules. The G-mic is a mixed modified amine system curing agent microcapsule, so the N-H bending vibration absorption peak is strong. The E-mic is an epoxy resin microcapsule, so it has an epoxy group absorption peak. In addition, there are strong vibrations near 1452 cm^{-1}, which is the asymmetric variable angle vibration absorption peak of CH_3 in the base asphalt and

E-mic, while its symmetrical bending vibration absorption peak is near 1347 cm^{-1}. The characteristic peak of the C-N stretching vibration is near 1256 cm^{-1} in the E-mic. The characteristic peak near 1094 cm^{-1} is the anti-symmetric stretching vibration absorption peak of C-C, and the characteristic absorption peak of the epoxy group is near 916 cm^{-1}. Therefore, no new characteristic peaks appeared after the E-mic was added to the base asphalt. It can be concluded that the E-mic was physically mixed with the asphalt, but there is no chemical reaction.

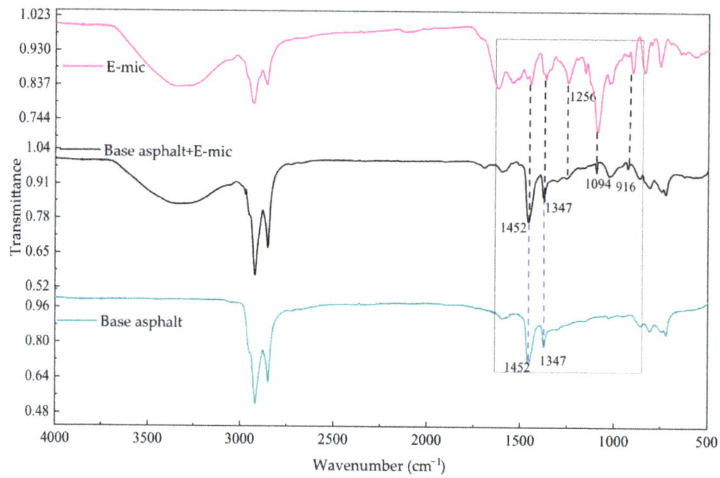

Figure 13. The infrared spectrum of the base asphalt and the base asphalt + E-mic.

3.2. *High-Temperature Rheological Analysis*

The ability of asphalt to resist permanent deformation at high temperatures has a very direct impact on the high-temperature rutting resistance of asphalt mixtures. SHRP proposed that the high-temperature performance of asphalt is characterized by the PG grading system, and the main evaluation indicator is the rutting factor ($G^*/\sin\delta$). The rutting factor changes with the temperature after the TS tests of the microcapsule/asphalt are shown in Figure 14. $G^*/\sin\delta$ represents the ability of asphalt to resist permanent deformation at high temperatures. That is to say, the larger the value of $G^*/\sin\delta$, the smaller the deformation of the asphalt under high temperatures and loads.

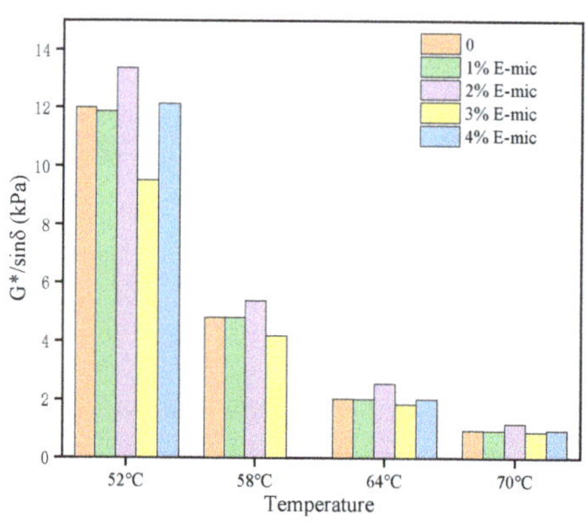

Figure 14. $G^*/\sin\delta$ of microcapsule/asphalt.

Figure 14 shows that, compared with that of the base asphalt, the G*/sin δ of the 2%E-mic is increased by 11.27%, 11.72%, 25.83, and 23.28%, respectively, at four temperatures. Compared with the base asphalt, the G*/sin δ of the 3%E-mic is reduced by 20.77%, 12.84%, 9.18%, and 5.62%, respectively, at four temperatures. The G*/sin δ of the 1% E-mic at all temperatures is smaller than that of the base asphalt, and the 4%E-mic is slightly larger than that of the base asphalt. The shell material of the E-mic is a thermosetting resin material, which is not sensitive to temperature and has a good elasticity. Therefore, the greater the dosage is, the greater the overall elasticity of the microcapsule asphalt material, while it is also affected by its liquid resin core material. Therefore, the greater the dosage of microcapsules, the greater the overall elasticity of the microcapsule/asphalt material will be. However, the overall mechanical properties of the microcapsules/asphalt are also influenced by the liquid epoxy resin core material. When the microcapsule/asphalt material is at a low temperature (52 °C), the elasticity of the 1% E-mic is enhanced, and the viscosity is reduced compared with the base asphalt, so the resistance to high-temperature deformation is reduced. As the dosage increases, the G*/sin δ reaches its peak at an E-mic dosage of 2% and its minimum at an E-mic dosage of 3%. When the microcapsule/asphalt material is at a higher temperature (70 °C), the G*/sin δ of the 1% E-mic, 3% E-mic, and 4% E-mic is less than 1.0 kPa. It can be speculated that the 2% E-mic and 4% E-mic have a better resistance to high-temperature deformation at the four temperatures.

3.3. Low-Temperature Crack Resistance Analysis

The low-temperature creep of microcapsule/asphalt tests was conducted by BBR and obtained the creep stiffness modulus (S) and its change rate (m) of different asphalt samples. A lower S and higher m indicate that the asphalt has good creep recovery characteristics at low temperatures. The test results are shown in Figure 15.

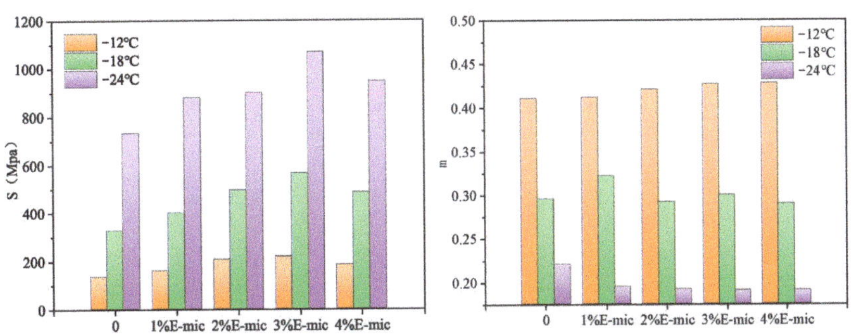

(a) Concentration change curve (b) Degradation rate change curve

Figure 15. The results of the low-temperature creep of microcapsule/asphalt tests.

Figure 15 shows that at different temperatures, the stiffness modulus of the microcapsule/asphalt first increases and then decreases with increasing dosages. It indicates that the addition of microcapsules could increase the stiffness modulus of the asphalt. The more microcapsules that are added, the more shell materials exist in the microcapsule/asphalt. Therefore, the brittleness of the microcapsule/asphalt material increases, S increases, and m decreases. When the dosage of the microcapsule reaches 3%, S reaches the maximum. At −12 °C, the low-temperature crack resistance and stress relaxation of all test specimens are better. The creep stiffness modulus and its change rate of all samples at −12 °C are plotted in Figure 16.

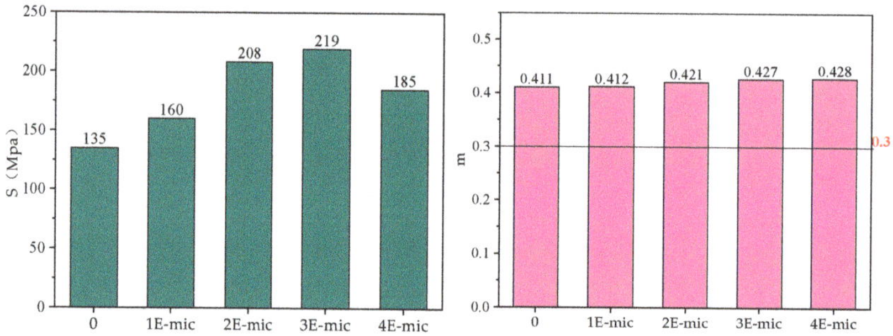

(**a**) Concentration change curve (**b**) Degradation rate change curve

Figure 16. The results of the low-temperature creep of microcapsule/asphalt tests at −12 °C.

As shown in Figure 16a, after adding the microcapsules, the stiffness modulus of the base asphalt was significantly increased. The stiffness modulus of microcapsule samples containing 1%, 2%, 3%, and 4% increased by 18.5%, 54.1%, 62.2%, and 37.0%, respectively. Although S increased by a large amount after the addition of the microcapsules, they still met the crack resistance requirements recommended by SHRP.

As shown in Figure 16b, the change rate of the stiffness modulus of the microcapsule samples containing 1%, 2%, 3%, and 4% increased by 0.2%, 2.4%, 3.9%, and 4.1%, respectively. After microcapsules were added, there was little change in m, and the m value of the modified asphalt was basically close to that of the base asphalt. Thus, it could be concluded that the addition of microcapsules did not significantly improve the relaxation performance of asphalt under low-temperature stress.

3.4. Fatigue Resistance Analysis

Each parameter of the LAS test corresponding to each sample was calculated based on the VECD theory.

(1) The Stress–Strain Analysis

The stress–strain curve during the LAS test is plotted in Figure 17. As can be seen from Figure 17, the base asphalt and microcapsule/asphalt demonstrate a yield phenomenon during fatigue loading. At low strain levels, the shear stress of all samples increases rapidly with the increase in the shear strain level. Afterwards, peaks appear in the range of the 8%~12% strain. The range of this region represents the sensitivity of the sample to strain under certain shear actions. The wider the peak area, the lower the sensitivity and the higher fatigue performance; otherwise, the fatigue performance of the asphalt sample decreases. When the shear strain increases and the shear stress decreases, the asphalt samples are considered to have begun to suffer fatigue failure. Within the peak area, stress is not sensitive to the change in strain. After the peak area, the stress of all the asphalt samples decreases rapidly with the increase in strain. Except that the width of the peak area of the 1% E-mic was the same as that of the base asphalt, the peak width of other samples basically increased with the increase in the microcapsule dosage. The order of the strain level corresponding to the maximum shear stress of each sample is the 4% E-mic, 2% E-mic, 1% E-mic, 3% E-mic, and base asphalt. The larger the yield strain of each sample indicates the stronger the anti-fatigue performance of the asphalt mixture. The results show that the yield stress of all microcapsule/asphalt samples is greater than that of the base asphalt. Based on the above analysis, therefore, the addition of microcapsules can effectively improve the fatigue performance of the base asphalt.

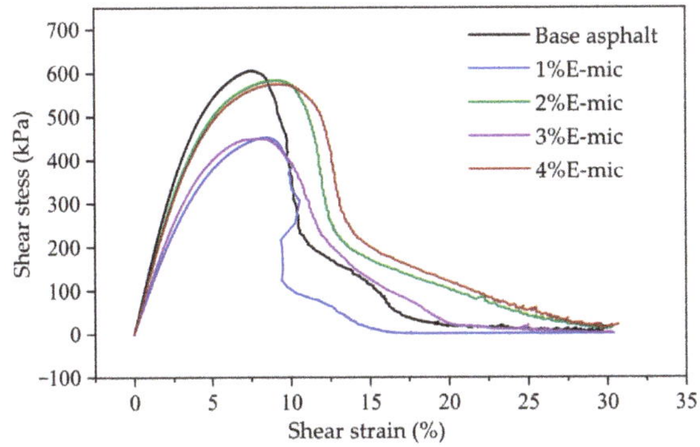

Figure 17. The stress–strain curve during the LAS test.

(2) Fatigue Damage Analysis

The fatigue damage characteristic curve of asphalt samples is plotted in Figure 18. As shown in Figure 18, when the damage intensity is less than 100, the cumulative fatigue damage (C) decreases rapidly. When the damage intensity (D) is in the range of 100–200, the change rate of C slows down and presents a basically linear decrease. Afterwards, when D is in the range of 200–300, the change rate of C flattens out. Finally, the change rate of C does not increase when D exceeds 300. The change rate of the damage parameter C of asphalt samples can reflect their fatigue life. The greater the rate, the shorter the fatigue life; otherwise, the fatigue life increases. When D is in the range of 100–200, except for the 1% E-mic, the change rate of C for other samples is lower than that of the base asphalt. That is, the corresponding fatigue life will be better.

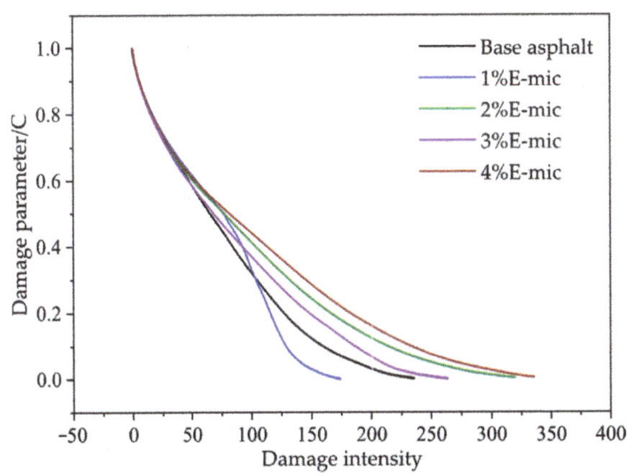

Figure 18. Fatigue damage curve.

(3) Fatigue Life Analysis

The influence of different strain levels on the fatigue performance of asphalt was analyzed and calculated. The fatigue life of the three strain levels, 2.5%, 5% and 10%, was calculated. The calculation results are shown in Figure 19.

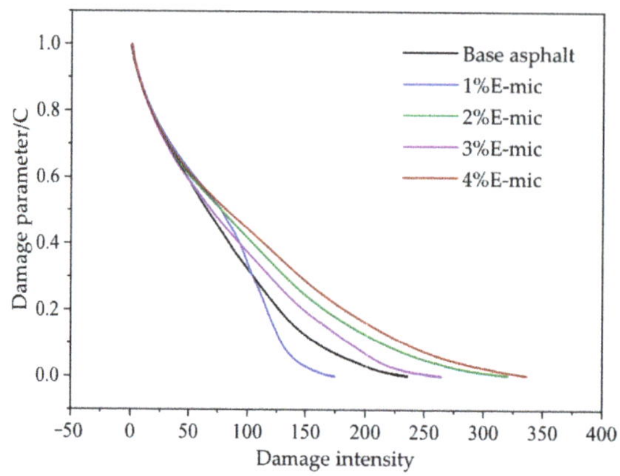

Figure 19. Fatigue life at different strain levels.

As the strain level increases, the fatigue life of asphalt samples significantly decreases. When the strain level is 2.5%, the fatigue life of the base asphalt is 1.2×10^5 times; when the strain level is 5%, the fatigue life is 968 times; and when the strain level increases to 10%, the fatigue life is only 76 times. When the strain level is 2.5%, the fatigue life of the 4% E-mic increases by 49.8% compared to that of the base asphalt. When the strain level is 5%, the fatigue life of the 2% E-mic and 4% E-mic significantly increased. Compared with the base asphalt, the fatigue life increases by 20.3% and 49.8%, respectively. When the strain level continues to increase to 10%, the fatigue life sequence is 4% E-mic > 2% E-mic > 3% E-mic > 1% E-mic > base asphalt. Therefore, microcapsules can effectively improve the fatigue performance of the base asphalt under high strain levels, and the 4% E-mic also has a better fatigue performance under low strain levels.

3.5. Self-Healing Property Analysis

The influence of adding microcapsules on the self-healing property of asphalt with fatigue damage was evaluated through the fatigue repair fatigue test. Figure 20 shows the test results with different dosages.

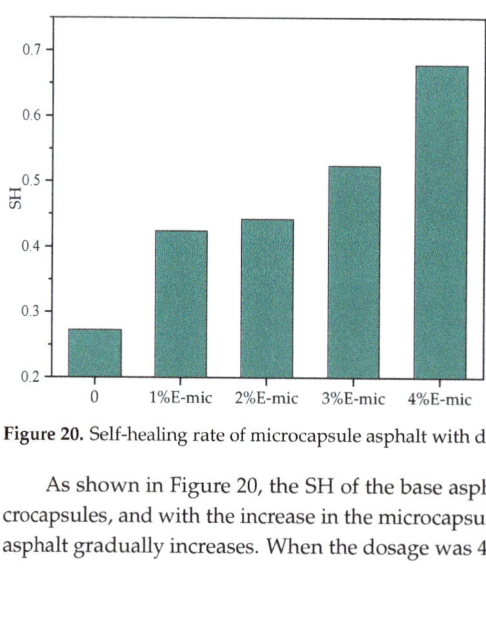

Figure 20. Self-healing rate of microcapsule asphalt with different dosages.

As shown in Figure 20, the SH of the base asphalt is significantly improved by microcapsules, and with the increase in the microcapsule dosage, the self-healing rate of the asphalt gradually increases. When the dosage was 4%, SH reached a maximum value of

67.8%, which is 149.2% higher than that of the base asphalt. The greater the dosage of the microcapsule, the better the self-healing effect of the asphalt. It can be observed that when micro-cracks appear in asphalt, the more microcapsules that are used, the more core material they release. Meanwhile, more cured epoxy resin that can repair the asphalt matrix will be produced. Therefore, with the increase in the microcapsule dosage, the self-healing performance of asphalt is significantly improved.

3.6. The Recommended Dosage of the Microcapsule

Based on the aforementioned analysis, the optimal dosage of microcapsules can be recommended. The TS test showed that the deformation resistance of microencapsulated modified asphalt was better at dosages of 2% and 4%. Low-temperature bending creep tests indicated that the stiffness modulus increased after the addition of the microcapsules but still met the crack resistance requirements recommended by SHRP. LAS tests showed that the microcapsules could effectively improve the resistance to fatigue of the base asphalt, and the resistance to fatigue was better when the dosage was 2% and 4%. According to the self-healing performance test, as the microcapsule content gradually increases, the self-healing performance of the asphalt is significantly improved, with a better dosage of 4%. The comparison results are shown in Table 5. In conclusion, considering the performance and production cost of the microcapsule-modified asphalt, it is recommended that the optimal dosage of microcapsules be 2%.

Table 5. Performance comparison of modified asphalt with different E-mic contents.

Content	Rutting Factor (64 °C)/kPa	Creep Stiffness Modulus (−12 °C)	Change Rate (−12 °C)	Self-Healing Rate
0	2.02	135	0.411	0.27
1	2.02	160	0.412	0.42
2	2.55	208	0.421	0.44
3	1.84	219	0.427	0.52
4	2.02	185	0.428	0.67

This study prepared two-component microcapsules (E-mic and G-mic) using modified epoxy resin E and curing agent G as core materials, with their basic properties (e.g., particle size distribution, thermal stability) characterized. On this basis, modified asphalt samples with varying microcapsule dosages were prepared, and their pavement performance and self-healing capability were systematically evaluated to determine the optimal dosage range. Existing research on microcapsule-modified asphalt has predominantly focused on asphalt's self-healing performance, with relatively little attention paid to the pavement performance of the modified asphalt. To address this gap, this study, following the successful preparation of weather-resistant two-component epoxy resin microcapsules, focuses on investigating the pavement performance of microcapsule-modified asphalt and reveals the influence of the microcapsule dosage on key asphalt properties, such as high-temperature resistance to permanent deformation and low-temperature rheological properties. The results indicate that the incorporation of two-component epoxy resin microcapsules enhances the asphalt's high-temperature performance, self-healing ability, and fatigue life, while exerting adverse effects on the low-temperature performance. These findings are applicable to two-component epoxy resin microcapsule-modified asphalt. Notably, this paper rarely elaborates on the mechanism underlying the performance enhancement of microcapsule-modified asphalt. Additionally, it has not yet addressed the performance evolution law of the modified asphalt after repair, nor has it explored the aging resistance of microcapsule-modified asphalt [33]. These are all areas that require further improvement in subsequent research.

4. Conclusions

Two-component microcapsules (E-mic and G-mic) were prepared in this study. Subsequently, the microcapsules prepared were added to the base asphalt to prepare the microcapsules/asphalt material. Afterwards, temperature scanning (TS) tests, low-temperature bending creep tests, and linear amplitude scanning (LAS) tests were conducted to evaluate the road performance of the microcapsule/asphalt. Finally, the self-healing property tests were performed to evaluate the self-healing property of the microcapsule/asphalt. The conclusions of this study are as follows.

1. For spherical micron-sized particles with a uniform shape and particle size, the average particle size of the E-mic was 23.582 μm, and that of the G-mic was 22.440 μm, showing a good normal distribution. In addition, they could maintain good stability at high temperatures.
2. The TS tests indicated that the rutting resistance of the base asphalt gradually decreases with the increase in the temperature, while the high-temperature deformation resistance showed a better performance when the microcapsule dosage was 2% and 4%.
3. The low-temperature bending creep stiffness test demonstrated that the addition of microcapsules increased the stiffness modulus of the base asphalt, but the creep rate was not obvious. At −12 °C, the low-temperature crack resistance and stress relaxation of all specimens were better. The microcapsule can slightly improve the stress relaxation at low temperatures.
4. The stress–strain curve of LAS shows that yield stress is greater than that of base asphalt except when the microcapsule dosage is 3%. The fatigue damage curve of the LAS shows that at the range of 100–200, the microcapsule/asphalt fatigue damage rate is lower than that of the base asphalt, except for the microcapsule dosage of 1%; that is, the corresponding fatigue life will be better. The microcapsules can effectively improve the anti-fatigue performance of the base asphalt at high strain levels, and microcapsules/asphalt with a microcapsule dosage of 4% also have a higher anti-fatigue performance at low strain levels.
5. Considering the high-temperature deformation resistance, the low-temperature crack resistance, the fatigue resistance, and the self-healing property of asphalt, the optimal dosage of the microcapsules is recommended to be 4%.

Author Contributions: Conceptualization, P.L. and C.Z.; methodology, R.J. and J.X.; formal analysis, X.S. and C.Z.; investigation, P.L.; data curation, R.J.; writing—original draft preparation, P.L.; writing—review and editing, R.J., J.X. and M.Z.; supervision, M.Z., P.L. and C.Z.; funding acquisition, M.Z., P.L. and C.Z. All authors have read and agreed to the published version of the manuscript.

Funding: This research was funded by the National Natural Science Foundation of China (Grant No.52378430) and the Guangdong Provincial Communication Group Technology Project (JT2023YB01).

Institutional Review Board Statement: Not applicable.

Informed Consent Statement: Not applicable.

Data Availability Statement: The original contributions presented in this study are included in the article. Further inquiries can be directed to the corresponding author.

Conflicts of Interest: Authors Pei Li and Chenlong Zhang were employed by the company Guangdong Road and Bridge Construction Development Co., Ltd. The remaining authors declare that the research was conducted in the absence of any commercial or financial relationships that could be construed as a potential conflict of interest.

Abbreviations

The following abbreviations are used in this manuscript:

E	Modified epoxy resin
G	Curing agent
MUF	Melamine modified urea–formaldehyde resin
E-mic	Modified epoxy resin microcapsule
G-mic	Curing agent microcapsule

References

1. Wang, L.; Cheng, P.; Zhao, Q. Research on Fatigue–Healing Performance of Asphalt Mixture Based on the Semicircular Bending Test. *Materials* **2023**, *16*, 6382. [CrossRef] [PubMed]
2. Ameri, M.; Sadeghiavaz, M. Microwave Healing Properties and Moisture Sensitivity of Asphalt Mixture Containing Iron Powder Filler. *Sci. Rep.* **2024**, *14*, 23154. [CrossRef] [PubMed]
3. Diao, H.; Ling, T.; Zhang, Z.; Peng, B.; Huang, Q. Multiscale Fatigue Performance Evaluation of Hydrated Lime and Basalt Fiber Modified Asphalt Mixture. *Materials* **2023**, *16*, 3608. [CrossRef] [PubMed]
4. Zheng, X.; Xu, W.; Cao, K.; Li, K. Self-Healing Behavior of Recycled Asphalt Prepared by Residue Oil of Straw Liquefaction Based on Molecular Dynamics Simulation. *Sci. Rep.* **2022**, *12*, 2718. [CrossRef] [PubMed]
5. Li, M.; Fini, E.; Jia, X.; Song, B.; Wang, Y. Effect of Aging on Healing Capacity of Bituminous Composites Containing Polyphosphoric Acid. *Materials* **2023**, *16*, 3333. [CrossRef] [PubMed]
6. Zhang, H.; Yao, T.; Cheng, F. Road Performance and Self-Healing Property of Bituminous Mixture Containing Urea–Formaldehyde Microcapsules. *Materials* **2024**, *17*, 943. [CrossRef] [PubMed]
7. Jin, J.; Miao, Y.; Zhao, H.; Chen, J.; Qing, L.; Mu, R.; Chen, X.; Li, Z. Study on the Self-Healing Performance of Microcapsules and Microcapsule-Containing Asphalt. *Sustainability* **2022**, *14*, 12231. [CrossRef]
8. Ding, X.; Cui, Y.; Ma, T.; Ye, F. Multiscale-Mechanical Analysis on Self-Healing Microcapsules Under Asphalt Pavement. *Constr. Build. Mater.* **2024**, *443*, 137670. [CrossRef]
9. Wang, Y.; Zhai, R.; Sun, B.; Liu, J.; Hao, P. Microcapsule Synthesis and Evaluation on Fatigue and Healing of Microcapsule-Based Asphalt by the Entropy and TOPSIS Method. *Int. J. Pavement Eng.* **2022**, *23*, 4610–4621. [CrossRef]
10. Zhang, H.; Wang, R.; Cheng, F. Study on the Properties and Self-Healing Ability of Asphalt and Asphalt Mixture Containing Methanol-Modified Melamine-Formaldehyde Microcapsules. *Mater. Today Commun.* **2024**, *41*, 111007. [CrossRef]
11. Xu, X.; Yuan, L.; Cong, P.; Wang, Z.; Guo, H.; Zhou, X.; Shao, T. Rheological and Healing Properties of Bitumen Containing Carbon Black Microcapsules. *Fuel* **2024**, *373*, 132399. [CrossRef]
12. Chen, B.; Liu, D.; Chen, A.; Wang, Y. Influence of Microcapsules on Self-Healing Performance and Road Performance of Dense-Graded Asphalt Mixture. *IOP Conf. Ser. Earth Environ. Sci.* **2021**, *768*, 12020. [CrossRef]
13. Xu, X.; Cong, P.; Wang, Z.; Zhang, T.; Zhou, X.; Yang, Q.; Zhao, Y.; Zhang, T. Application of Microencapsulated Waste Cooking Oil in Bitumen: Responsiveness and Bitumen Performance Changes. *Fuel* **2025**, *402*, 136026. [CrossRef]
14. Lu, T.; Hofko, B.; Sun, D.; Mirwald, J.; Eberhardsteiner, L.; Hu, M. Microscopic and Rheologic Characterization of Third Generation Self-Repairing Microcapsule Modified Asphalt. *Constr. Build. Mater.* **2023**, *400*, 132841. [CrossRef]
15. Wen, Y.; Ma, F.; Fu, Z.; Li, C.; Shi, K.; Dai, J.; Zhu, C. Evaluation of the Microcapsules on the Rheological and Self-Healing Performance of Asphalt. *Constr. Build. Mater.* **2023**, *409*, 133982. [CrossRef]
16. Yang, S.; Ji, J.; Tao, H.; Muhammad, Y.; Huang, J.; Wang, S.; Wei, Y.; Li, J. Fabrication of Urea Formaldehyde–Epoxy Resin Microcapsules for the Preparation of High Self-Healing Ability Containing SBS Modified Asphalt. *Polym. Compos.* **2021**, *42*, 4128–4137. [CrossRef]
17. Wang, Y.-Y.; Tan, Y.-Q.; Lv, H.-J.; Han, M.-Z. Evaluation of Rheological and Self-Healing Properties of Asphalt Containing Microcapsules Modified with Graphene. *Constr. Build. Mater.* **2022**, *357*, 129287. [CrossRef]
18. Ma, E.; Chen, X.; Lai, J.; Kong, X.; Guo, C. Self-Healing of Microcapsule-Based Materials for Highway Construction: A Review. *J. Traffic Transp. Eng. (Engl. Ed.)* **2023**, *10*, 368–384. [CrossRef]
19. Wen, Y.; Ma, F.; Fu, Z.; Dong, W.; Jia, M.; Dai, J.; Hou, Y.; Jiang, X.; Hao, Y. Influence of Hybrid Microcapsules on the Low-Temperature Cracking Resistance of Asphalt Binder. *Constr. Build. Mater.* **2025**, *480*, 141525. [CrossRef]
20. Ren, Y.; Hao, P. Modification Mechanism and Enhanced Low-Temperature Performance of Asphalt Mixtures with Graphene-Modified Phase-Change Microcapsules. *Constr. Build. Mater.* **2022**, *320*, 126301. [CrossRef]
21. Tao, M.; Xu, T. Development and Properties of Bio-Based Rice Oil Microcapsule and Its Dispersibility in Emulsified Asphalt. *J. Ind. Eng. Chem.* **2024**, *131*, 503–513. [CrossRef]
22. Xu, X.; Yuan, L.; Cong, P.; Wang, Z.; Zhou, X.; Wang, J.; Liu, J. Self-Healing Microcapsule Properties Improvement Technology: Key Challenges and Solutions for Application in Asphalt Materials. *Constr. Build. Mater.* **2024**, *439*, 137298. [CrossRef]

23. Tian, Y.; Zheng, M.; Li, P.; Zhang, J.; Qiao, R.; Cheng, C.; Xu, H. Preparation and Characterization of Self-Healing Microcapsules of Asphalt. *Constr. Build. Mater.* **2020**, *263*, 120174. [CrossRef]
24. Pei, Z.; Yi, J.; Xu, M.; Ai, X.; Cao, J.; Hu, W.; Gao, L.; Feng, D. Exploration of the Design Theory of Crack-Resistant Rejuvenator for Warm-Mix Recycled Asphalt Mixtures with High RAP Contents. *J. Clean. Prod.* **2023**, *388*, 135855. [CrossRef]
25. Chen, S.; Liu, Q.; Bi, Y.; Yu, B.; Zhang, J. Self-Healing Effect of Various Capsule-Core Materials on Asphalt Materials. *Adv. Civ. Eng.* **2022**, *2022*, 5372501. [CrossRef]
26. Norambuena-Contreras, J.; Arteaga-Perez, L.E.; Guadarrama-Lezama, A.Y.; Briones, R.; Vivanco, J.F.; Gonzalez-Torre, I. Microencapsulated Bio-Based Rejuvenators for the Self-Healing of Bituminous Materials. *Materials* **2020**, *13*, 1446. [CrossRef] [PubMed]
27. Zhai, Z.; Li, H.; Feng, L.; Li, F. Effect of UV Light on the Barrier Performance of Aluminum Powder/Epoxy Coating. *Polymers* **2022**, *14*, 2405. [CrossRef] [PubMed]
28. Yalikun, N.; Yu, S.; Yang, H.; Liu, C.; Zhang, H.; Wang, Q.; Zhang, Y.; Chen, F. Preparation and Performance Study of Oxidized CR/SBS Composite Modified Asphalt. *J. Environ. Chem. Eng.* **2025**, *13*, 117094. [CrossRef]
29. Wen, D.; Ge, D.; Wang, Y.; Lv, S.; Liu, Q.; Liu, S. Effectiveness of Different Categories of Light Oils in Partially Reactive Crumb Rubber-Modified Asphalt. *Materials* **2025**, *18*, 1871. [CrossRef] [PubMed]
30. Chen, C.; Ma, S.; Zhu, S.; Fan, L.; Liu, P.; Lin, J.; Li, Y. Comparison and Analysis of Testing Methods for Storage Stability of Sbs Modified Asphalt. *Highw. Eng.* **2023**, *48*, 173–178.
31. *AASHTO TP 101-14*; Standard Method of Test for Estimating Damage Tolerance of Asphalt Binders Using the Linear Amplitude Sweep. American Association of State Highway and Transportation Officials: Washington, DC, USA, 2014.
32. Sun, D.; Pang, Q.; Zhu, X.; Tian, Y.; Lu, T.; Yang, Y. Enhanced Self-Healing Process of Sustainable Asphalt Materials Containing Microcapsules. *ACS Sustain. Chem. Eng.* **2017**, *5*, 9881–9893. [CrossRef]
33. Ābele, A.; Merijs-Meri, R.; Bērziņa, R.; Zicāns, J.; Haritonovs, V.; Ivanova, T. Effect of Bio-Oil on Rheological and Calorimetric Properties of RTFOT Aged Bituminous Compositions. *Int. J. Pavement Res. Technol.* **2021**, *14*, 537–542. [CrossRef]

materials

MDPI

Article

A Study on the Characterization of Asphalt Plant Reclaimed Powder Using Fourier Transform Infrared Spectroscopy

Hao Wu [1], Daoan Yu [2], Wentao Wang [1], Chuanqi Yan [2], Rui Xiao [3], Rong Chen [3], Peng Zhang [2,*] and Hengji Zhang [4]

1 Shandong Provincial Road and Bridge Group Co., Ltd., Jinan 250014, China
2 School of Civil Engineering, Southwest Jiaotong University, Chengdu 610031, China; ycq@swjtu.edu.cn (C.Y.)
3 Sichuan Transportation Construction Group Co., Ltd., Chengdu 610041, China
4 Shanxi Transportation Holdings Science & Technology Transformation Co., Ltd., Taiyuan 030012, China
* Correspondence: pz9116263@gmail.com

Abstract

Asphalt plant reclaimed powder is a common solid waste in road engineering. Reusing reclaimed powder as filler holds significant importance for environmental protection and resource conservation. The key factors affecting the feasibility of reclaimed powder reuse are its acidity/alkalinity and cleanliness. Traditional evaluation methods, such as the methylene blue test and plasticity index, can assess reclaimed powder properties to guide its recycling. However, these methods suffer from inefficiency, strong empirical dependence, and high variability. To address these limitations, this study proposes a rapid and precise evaluation method for reclaimed powder properties based on Fourier transform infrared spectroscopy (FTIR). To do so, five field-collected reclaimed powder samples and four artificial samples were evaluated. Scanning electron microscopy (SEM), X-ray fluorescence spectroscopy (XRF), and X-ray diffraction (XRD) were employed to characterize their microphase morphology, chemical composition, and crystal structure, respectively. Subsequently, FTIR was used to establish correlations between key acidity/alkalinity, cleanliness, and multiple characteristic peak intensities. Representative infrared characteristic peaks were selected, and a quantitative functional group index (Is) was proposed to simultaneously evaluate acidity/alkalinity and cleanliness. The results indicate that reclaimed powder primarily consists of tiny, crushed stone particles and dust, with significant variations in crystal structure and chemical composition, including calcium carbonate, silicon oxide, iron oxide, and aluminum oxide. Some samples also contained clay, which critically influenced the reclaimed powder properties. Since both filler acidity/alkalinity and cleanliness are affected by clay (silicon/carbon ratio determining acidity/alkalinity and aluminosilicate content affecting cleanliness), this study calculated four functional group indices based on FTIR absorption peaks, namely the Si-O-Si stretching vibration (1000 cm^{-1}) and the CO_3^{2-} asymmetric stretching vibration (1400 cm^{-1}). These indices were correlated with conventional testing results (XRF for acidity/alkalinity, methylene blue value, and pull-off strength for cleanliness). The results show that the Is index exhibited strong correlations (R^2 = 0.89 with XRF, R^2 = 0.80 with methylene blue value, and R^2 = 0.96 with pull-off strength), demonstrating its effectiveness in predicting both acidity/alkalinity and cleanliness. The developed method enhances reclaimed powder detection efficiency and facilitates high-value recycling in road engineering applications.

Keywords: reclaimed powder; infrared spectroscopy (FTIR); acidity/alkalinity; cleanliness

1. Introduction

During asphalt mixture production, raw materials in asphalt plants undergo transportation, production, heating, and mixing processes. Due to the hot-mix nature of asphalt mixtures, a significant amount of heavily polluted dust particles is generated [1]. These particles are collected by the plant's dust removal system and stored in designated containers. Typically, these particles have a size smaller than 0.075 mm and are referred to as reclaimed powder (or reclaimed dust) [2].

The current practice of stockpiling or landfilling reclaimed powder not only occupies valuable land resources but also reduces land-use efficiency while creating negative impacts on energy production, resource utilization, and environmental protection [3]. With the rapid expansion of highway infrastructure [4], failure to properly recycle this material poses significant challenges to sustainable development [5].

Ratnasamy Muniandy [6] investigated the influence of mineral powder on the performance of stone mastic asphalt (SMA). The study revealed that excessive use of powder significantly increases the stiffness of the mastic portion. However, when key parameters of powder—including density, gradation distribution, appearance, the hydrophilicity coefficient, and plasticity index—meet specified requirements, it can be effectively utilized in SMA applications.

Chi-Wei Chen's research [7] demonstrated that reclaimed powder not only satisfies the quality standards for mineral powder but also exhibits stronger alkalinity, which enhances the bonding between asphalt and aggregates. The Pennsylvania Department of Transportation [8] conducted a study on six types of reclaimed powder, concluding that when the dust content is maintained below 2%, the impact on mixture performance remains negligible.

When utilized as a filler replacement, reclaimed powder must meet stringent requirements regarding its acidity/alkalinity and cleanliness. Research by Najib Mukhtar et al. [9] demonstrated that the pH characteristics of fillers significantly influence the adhesion between filler particles and asphalt binder in asphalt mixtures. Frederick M. Fowkes [10] further established that this interfacial bonding strength shows strong correlation with the material's silica content, which serves as a key indicator of its acid–base properties.

The cleanliness of reclaimed powder, particularly its clay content, represents another crucial quality parameter. Wu Xuan et al. [11] found that impurities in filler materials can adversely affect the high-temperature performance of asphalt mixtures. Complementary research by Gajewski [12] revealed that excessive impurities may increase the optimum water content, leading to reduced binder viscosity [13], consequently impairing both the mechanical properties and long-term durability of the mixture.

Current evaluation methods for these parameters are specified in the National Standard "Rock Classification and Nomenclature Scheme". The standard recommends X-ray fluorescence (XRF) spectroscopy for determining filler acidity/alkalinity, while cleanliness assessment primarily focuses on clay content through either methylene blue value (MBV) testing or plasticity index (PI) measurement [14].

Although these methods provide feasibility for testing reclaimed powder properties, they still present several problems. Microanalysis equipment like XRF is difficult to widely implement during construction processes; while methylene blue and plasticity index tests are relatively common, their testing procedures are overly complex, heavily empirical, require highly skilled operators, and show significant variability. These issues limit the large-scale, high-efficiency recycling applications of reclaimed powder. Hou [15] used XRF, noting that while XRF instruments have high sensitivity, their high cost and bulk size make them unsuitable for portable use. JM Moreno-Maroto [16] conducted an examination of fine-grained soil classification systems based on plasticity, with test results showing that plasticity index tests have clear deficiencies when testing particles of different minerals, requiring comprehensive judgment through other mineral composition analysis tests. SK Singh [17] and others researched detection methods for material clay content, with studies showing that the current standard methylene blue test method relies on assumed preconditions, making it quite empirical.

Currently, many field laboratories of China are equipped with portable infrared spectrometers for measuring SBS content [18]. These infrared spectrometers can rapidly detect the chemical composition of various materials, including fillers and reclaimed powder [19]. At the same time, the operation is simple and easy to carry out, which provides a hardware foundation for the rapid and accurate testing of the acidity/alkalinity and cleanliness of the reclaimed powder [20]. Given these circumstances, this paper proposes a new method for evaluating and classifying reclaimed powder from asphalt mixing plants based on Fourier transform infrared spectroscopy (FTIR) tests. This study uses infrared spectrometers to scan reclaimed powder, extract relevant functional group information, establishes correlations between this information and the acidity/alkalinity and cleanliness indicators of reclaimed powder, and ultimately guides the classification and recycling of reclaimed powder.

2. Materials and Methods

2.1. Materials

To develop a Fourier transform infrared spectroscopy (FTIR)-based detection method for reclaimed powder, five types of reclaimed powder were collected from different fields, as shown in Figure 1. During collection, the damper opening of the mixing plant's dust removal system was set to 32%, with a flue gas flow velocity of 13.8 m/s. For convenience in subsequent descriptions, the reclaimed powder samples are abbreviated as RP1 to RP5.

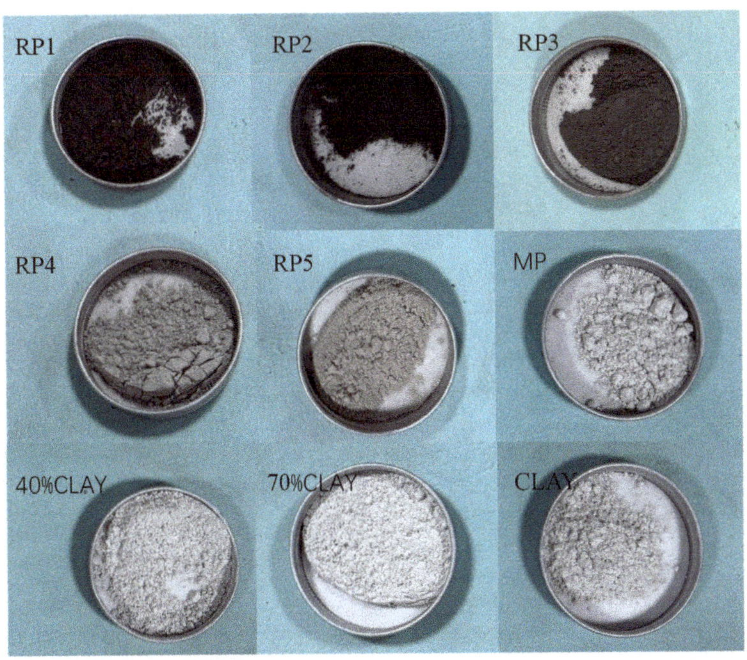

Figure 1. The powders used in this study.

Given the significant variability in field-collected reclaimed powder, artificial reclaimed powder was prepared to obtain more trend-conclusive results. The artificial samples were created by artificially adding 40% and 70% clay to limestone mineral powder (MP), designated as 40% clay and 70% clay, respectively, to simulate different cleanliness levels. Additionally, pure clay was also tested for reference.

The basic properties of the nine powder materials were tested, and the results are shown in Table 1. The density and particle size distribution of the reclaimed powder were similar to those of mineral powder, but significant color variations were observed among the reclaimed powder samples. These color differences may be attributed to substantial variations in the mineral composition of aggregates used in different regions.

Table 1. Basic parameters of reclaimed powders.

Property	Test Method	RP1	RP2	RP3	RP4	RP5	MP	40% CLAY	70% CLAY	CLAY
Density (g/cm^3)	JTG 3432-2024	2.42	2.51	2.13	2.39	2.33	2.41	2.43	2.44	2.47
Particle size (nm)	Laster particle test	2946	3237	2891	2973	2976	3019	2985	3022	2926
Color	\	black	black	black	white	white	brown	brown	brown	brown
Source		municipal	municipal	municipal	highway	highway	lab-made	lab-made	lab-made	lab-made

2.2. Methods

2.2.1. Scanning Electron Microscopy (SEM)

To observe the micromorphology of the powder materials, scanning electron microscopy (SEM) analysis was conducted. SEM testing was performed using a JSM-IT500 instrument (JEOL, Kyoto, Japan) with an acceleration voltage of 20 kV. Prior to testing, the samples underwent a 5 min gold sputtering process under vacuum to enhance surface conductivity. This pretreatment prevents charge accumulation effects that could otherwise interfere with electron signal transmission and lead to image distortion, deformation, or instability.

2.2.2. X-Ray Fluorescence Spectroscopy (XRF) Test

To determine the elemental composition of the reclaimed powder and subsequently evaluate its acidity/alkalinity, X-ray fluorescence (XRF) spectroscopy was performed on all nine powder samples. The XRF testing was conducted using a Shimadzu XRF-1800 spectrometer (Shimadzu, Kyoto, Japan) with the following specifications: rhodium target X-ray tube; maximum power output: 4 kW; window thickness: 75 μm; elemental detection range: oxygen (O) to uranium (U); and analysis mode: top irradiation geometry.

2.2.3. X-Ray Diffraction (XRD) Test

X-ray diffraction (XRD) analysis was conducted to characterize the phase composition of the reclaimed powder and identify its crystalline species and structure. The XRD testing was performed using a Bruker D8 Advance diffractometer (Rigaku Corporation, Kyoto, Japan) with the following experimental parameters: scanning range: 10° to 90° (2θ); scanning speed: 2°/min; and step size: 0.02°.

2.2.4. Fourier Transform Infrared Spectroscopy (FTIR) Test

Fourier transform infrared spectroscopy (FTIR) was employed to analyze the powder samples. The infrared spectrometer model was Nicolet iS20 FTIR (Thermo Fisher Scientific, Waltham, MA, USA). Using attenuated total reflection (ATR) accessories, the apodization function was Happ-Genzel, the number of scans was 64, and FTIR scanning was performed in the wavenumber range of 400–2000 cm^{-1}.

Approximately 1 g of the sample was placed on the ATR crystal surface for spectral acquisition. The obtained spectra were subsequently processed using MATLAB software (R2024b) for quantitative analysis, from which evaluation indices for both acidity/alkalinity and cleanliness were extracted using portable infrared spectrometers, as shown in Figure 2.

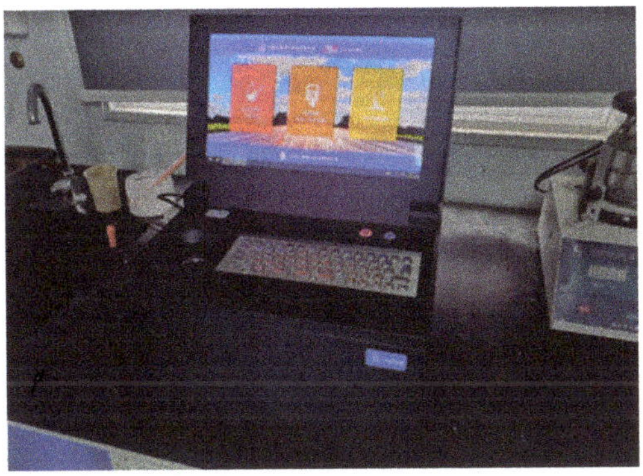

Figure 2. Portable infrared spectrometers.

2.2.5. Methylene Blue (MB) Test

To evaluate the cleanliness of the reclaimed powder, methylene blue tests were conducted on the powder materials in accordance with the JTG 3432-2024 (Test Methods of Aggregates for Highway Engineering, Ministry of Transport of the People's Republic of China, 2024) [21]. The standard methylene blue solution was prepared using analytically pure methylene blue crystals and distilled water, with titration performed on the suspension using quantitative filter paper. The test results were expressed as the methylene blue value (MBV) of the materials. The MBV primarily correlates with the content of fine particles having large specific surface areas (such as clay) in the filler.

3. Results and Discussion

3.1. SEM Observation Results (Filler Micromorphology)

To observe the micromorphology of various reclaimed powder samples, SEM scanning was performed to capture microscopic images under different magnification conditions. The SEM results for different reclaimed powders are presented in Figures 3–7. The left panel of each figure shows the SEM image at 50×, while the right panel displays a locally magnified view at 2000×, with the scale bar positioned at the bottom left corner.

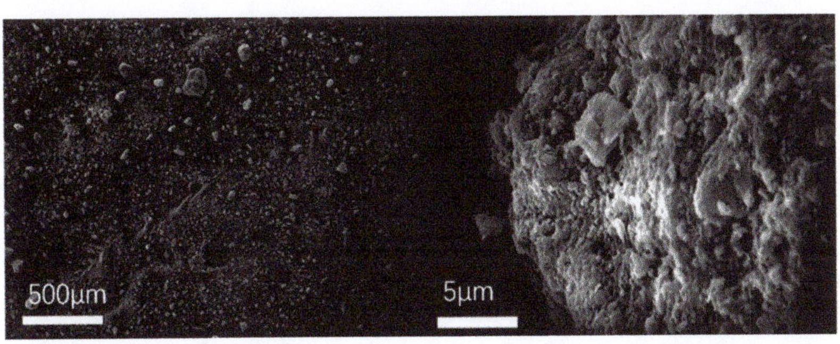

Figure 3. SEM micrograph of RP1.

Figure 4. SEM micrograph of RP2.

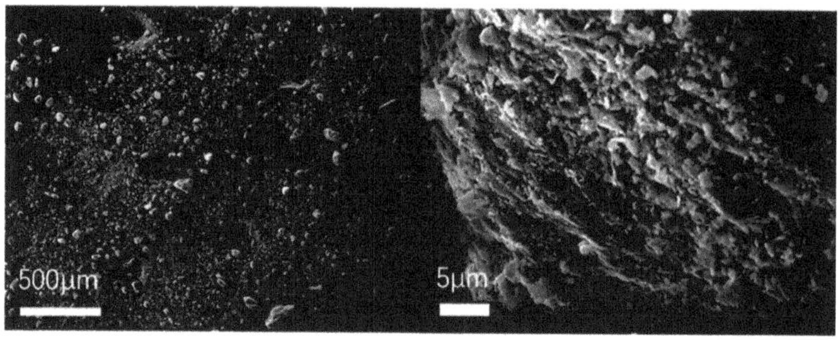

Figure 5. SEM micrograph of RP3.

Figure 6. SEM micrograph of RP4.

Figure 7. SEM micrograph of RP5.

The images reveal that the reclaimed powder consists of accumulated fine particles approximately 1 μm in size. Most particles exhibit irregular flaky and flocculent morphologies, with a minority showing porous structures and spherical shapes. According to research by HE Bergna [22], the spherical particles are primarily composed of silicon oxide. Overall, the reclaimed powder demonstrates small particle sizes, significant variability, and distinct micromorphological differences across samples from different regions.

Based on the sources of the reclaimed powder, its primary components were determined to be mineral powder and dust particles. To better characterize the micromorphological features of the reclaimed powder, supplementary SEM images of mineral powder and soil were obtained for comparison (Figures 8 and 9).

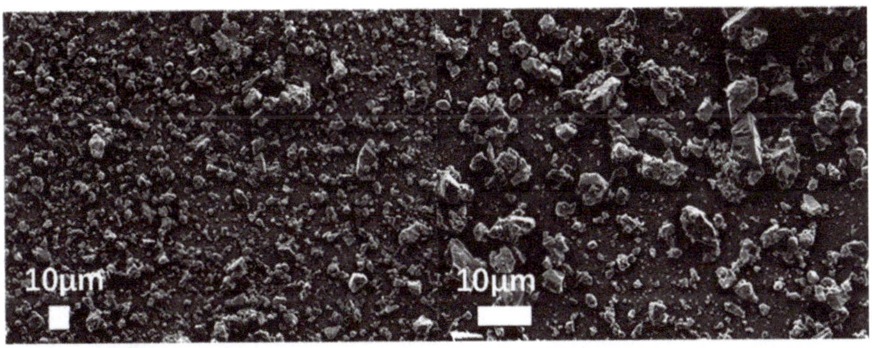

Figure 8. SEM micrograph of MP.

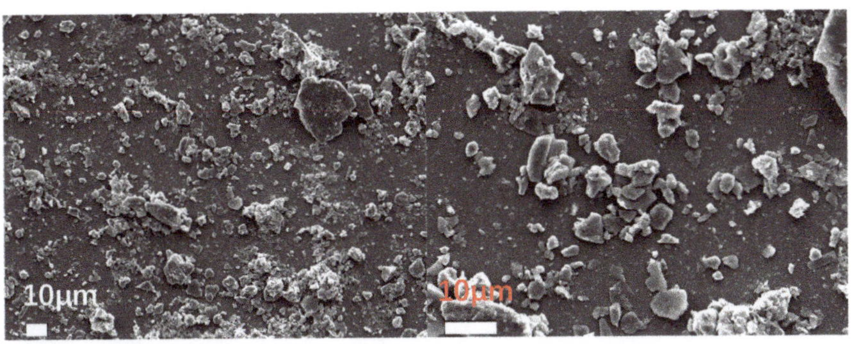

Figure 9. SEM micrograph of soil.

The mineral powder exhibited more uniform morphology with larger particles, while the soil particles appeared more irregular—a characteristic potentially attributable to the presence of organic matter in the soil. Comparative analysis revealed that the reclaimed powder shared morphological similarities with both reference materials, suggesting its composition likely consists of fragmented rock particles and dust.

3.2. *XRF Test Results (Filler Acidity/Alkalinity)*

XRF testing was conducted to determine the compound contents in the nine powder materials, with the measured compound contents shown in Table 2. The data for mineral powder, 40% clay, 70% clay, and pure clay showed linear correlations. RP1, RP2, and RP3 contained relatively higher contents of calcium carbonate ($CaCO_3$), silicon oxide (SiO_2), and iron oxide (Fe_2O_3), along with minor amounts of other compounds including aluminum oxide (Al_2O_3) and potassium oxide (K_2O). In contrast, RP4 and RP5 were predominantly composed of calcium carbonate.

Table 2. Chemical composition of nine powder materials (%).

Sample	SiO_2	$CaCO_3$	Al_2O_3	Fe_2O_3	MgO	TiO_2	K_2O	Total
RP1	23	41	3	21	1	4	4	98
RP2	21	22	6	39	1	7	2	98
RP3	22	17	6	42	2	6	3	98
RP4	1	97	0	1	0	0	0	99
RP5	2	90	1	2	4	0	1	100
MP	0	99	0	0	0	0	0	100
40% CLAY	18	74	3	2	1	0	0	100
70% CLAY	46	37	9	5	1	0	1	99
CLAY	73	5	12	6	1	0	2	99

The compound contents of the nine powder materials are shown in Figure 10. It can be observed that RP4, RP5, and mineral powder have relatively simple compositions, consisting mainly of calcium carbonate ($CaCO_3$), while RP1, RP2, and RP3 contain higher amounts of impurities such as silicon oxide (SiO_2) and iron oxide (Fe_2O_3). These results demonstrate significant variability in the properties of reclaimed powder, indicating the necessity of property testing prior to use. RP1, RP2, and RP3 were collected from municipal projects, whereas RP4 and RP5 came from highway projects. This indicates that construction quality control also plays an important role in ensuring the property consistency of reclaimed powder.

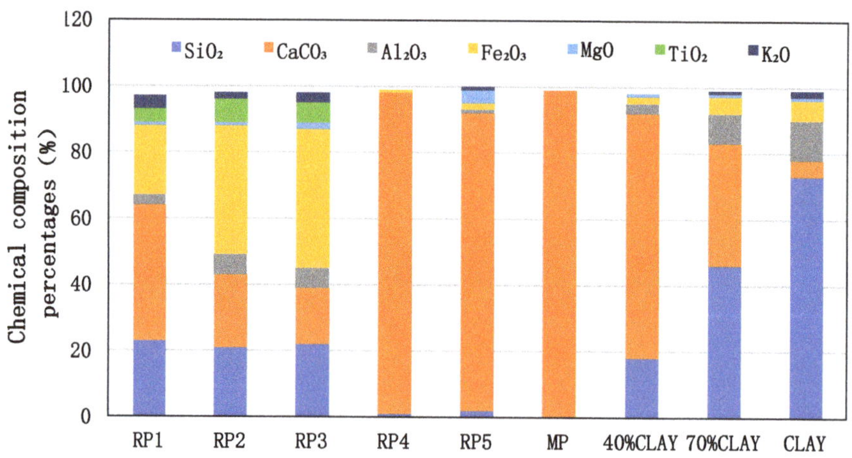

Figure 10. XRF results of chemical composition for nine powders.

On the other hand, the silicon oxide (SiO_2) content in mineral powder, 40% clay, 70% clay, and pure clay showed a linear increase, which was directly proportional to the clay content in these four materials. The SiO_2 content exceeded 70% in pure clay. Higher SiO_2 content in the powder materials resulted in lower alkalinity. It can be concluded that the addition of clay not only increases the acidity of the filler but also reduces its cleanliness. Therefore, for reclaimed powder, special attention must be paid to controlling the clay content. In general, when selecting aggregates, aggregates with high calcium carbonate content are called alkaline aggregates, and aggregates with high silica content are called acidic aggregates. The interaction between calcium carbonate and asphalt is stronger than that of silicon oxide, because calcium carbonate is a polar molecule, while silicon oxide crystal is a symmetrical atomic crystal composed of atoms. Therefore, when interacting with polar polymer asphalt, polar calcium carbonate exhibits stronger interaction [23,24]. This also explains the phenomenon that an alkaline aggregate has good adhesion to asphalt.

Since clay primarily consists of aluminosilicates, subsequent FTIR analysis may utilize silicon-related characteristic peaks (such as the Si-O-Si absorption peak at 1000 cm^{-1}) to characterize clay presence and evaluate both the acidity/alkalinity and cleanliness of the filler materials.

3.3. XRD Test Results (Filler Crystal Structure)

X-ray diffraction (XRD) analysis was conducted to examine the crystallization characteristics and crystal types of compounds in the nine powder materials, with the results shown in Figure 11. The diffraction intensities of various crystals exhibited a linear distribution that corresponded well to the formulation ratios of the four material types. The mineral powder displayed single absorption peaks, indicating its simple composition, while more complex waveforms suggested higher impurity content in other samples.

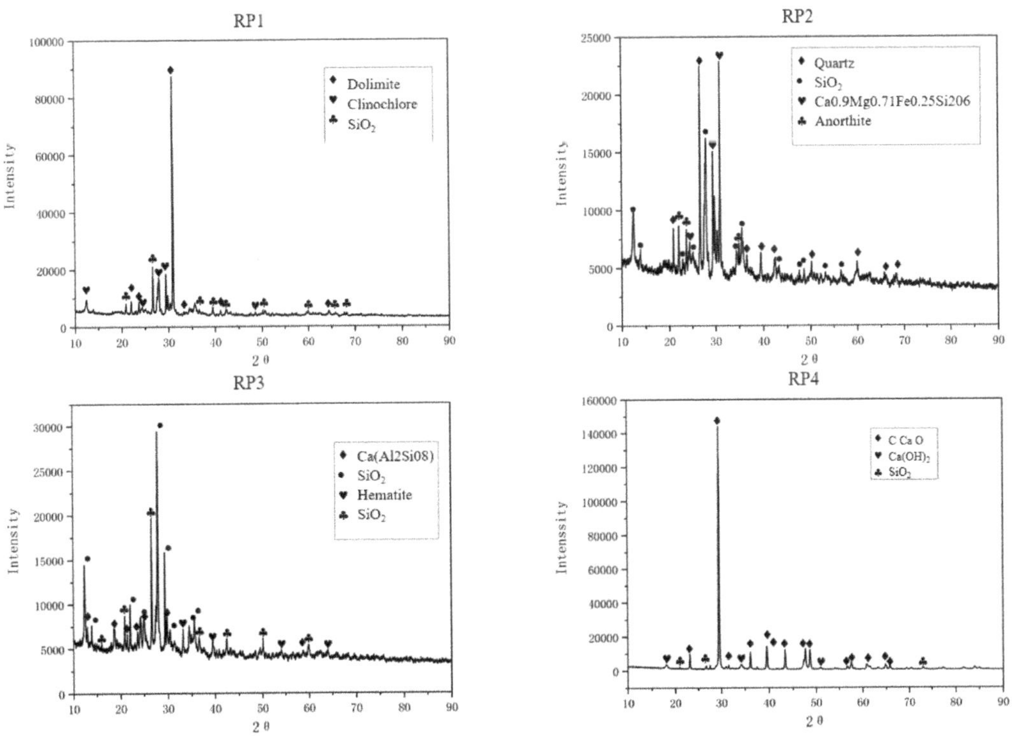

Figure 11. *Cont.*

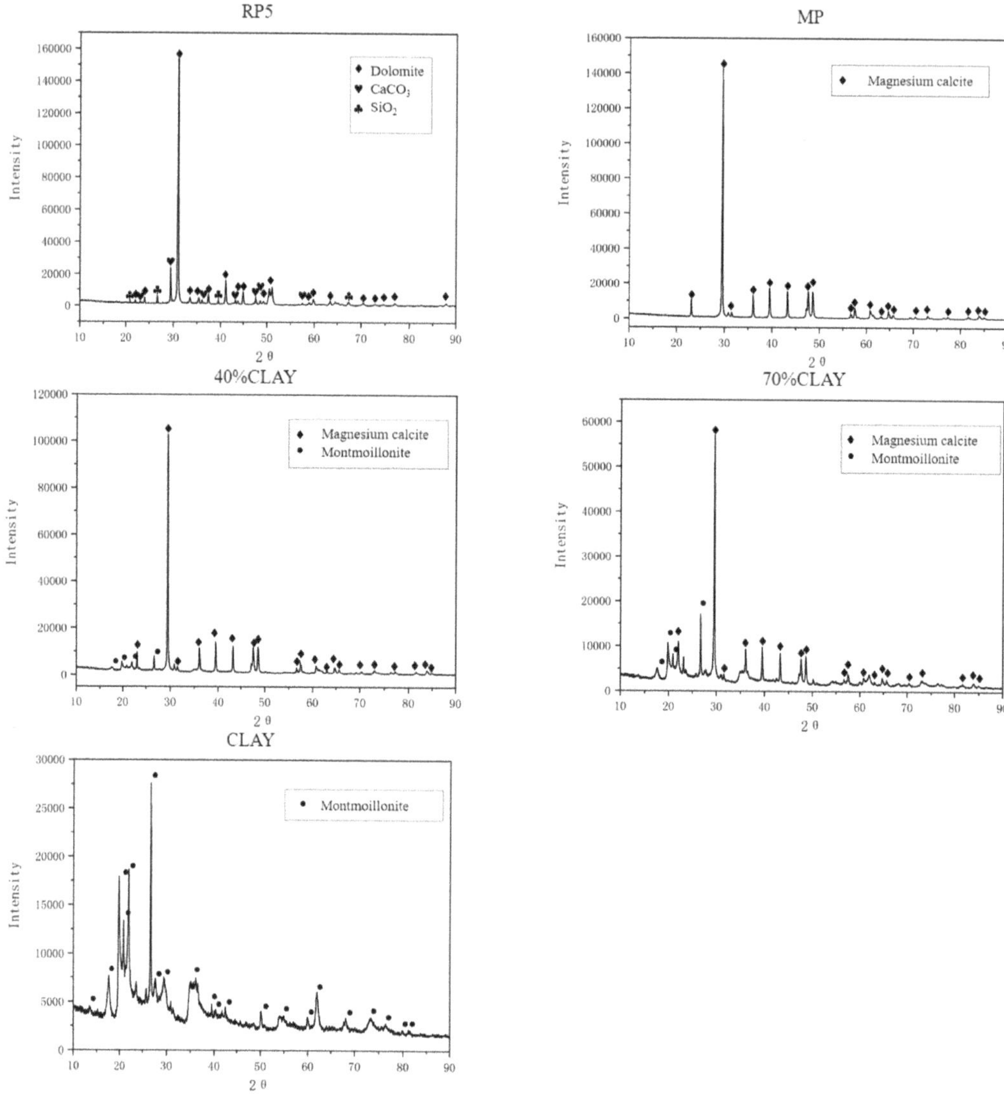

Figure 11. XRD spectra results of nine powders.

Peak fitting analysis revealed that the mineral powder had a relatively simple crystalline composition, predominantly calcite (primary component: calcium carbonate, $CaCO_3$). As the clay content increased, the spectra became progressively more complex with additional impurity peaks, indicating higher clay impurity content primarily composed of aluminosilicates. All five reclaimed powder samples exhibited sharp characteristic peaks, demonstrating their highly crystalline nature. The main crystalline phases in the reclaimed powder were calcium carbonate ($CaCO_3$) and silicon oxide (SiO_2) crystals, including accessory minerals such as anorthite, quartz, diopside, and chlorite, along with various metal oxides. Compared to RP4 and RP5, samples RP1, RP2, and RP3 showed more impurity peaks, reflecting their more complex composition and higher impurity content, which is consistent with the XRF experimental results.

3.4. Methylene Blue Test Results (Filler Cleanliness)

The methylene blue (MB) test was employed to evaluate the cleanliness of fillers, primarily reflecting the content of fine particles with large specific surface areas (e.g., clay) in the filler materials. A higher MB value (MBV) indicates poorer cleanliness, while a lower MBV suggests cleaner material. To investigate the cleanliness of the reclaimed powder, MB tests were conducted on all nine powder samples, with the results presented in Table 3.

Table 3. Methylene blue test results of nine powders.

Test Material	MBV
RP1	1.25
RP2	10
RP3	4
RP4	1
RP5	1.25
MP	0.5
40% CLAY	138.5
70% CLAY	241.25
CLAY	344.25

The MBVs of mineral powder, 40% clay, 70% clay, and pure clay exhibited a linear relationship, confirming the significant influence of clay content on material cleanliness. Due to substantial variations in particle composition and impurity levels among reclaimed powders from different sources, their MBVs showed considerable differences. RP4 and RP5 demonstrated better cleanliness, whereas RP1, RP2, and RP3 had higher MBVs, highlighting the importance of construction quality control.

Notably, RP4 showed a similar MBV to mineral powder, which can be combined with previous test results indicating their analogous calcium carbonate ($CaCO_3$) composition. This suggests that RP4 could potentially be a substitute for mineral powder in recycling applications, despite minor micromorphological differences.

3.5. FTIR Spectroscopy Analysis

3.5.1. Functional Group Analysis and Characteristic Peak Index Calculation of Fillers

The XRD and XRF instruments mentioned above are primarily research equipment and are hard to access in a field lab, whereas infrared spectroscopy, as a more common testing instrument, can be equipped in field laboratories. FTIR spectroscopy offers multiple advantages for powder analysis including a fast scanning speed, simple procedures, and accurate results. Therefore, this study proposes using functional groups in infrared spectra as indicators to evaluate both the cleanliness and acidity/alkalinity of powder materials. The preliminary SEM and XRD characterizations provide a morphological reference for the recycled powder, which is beneficial to determine the appropriate FTIR scanning range and characteristic peak selection. To analyze the functional groups related to these properties in the reclaimed powder, FTIR scans were performed on all nine samples. Each homogenized sample underwent triplicate tests under identical conditions to verify spectral reproducibility, with results showing high consistency across replicates. Since no absorption peaks were observed at wavenumbers above 2000 cm^{-1}, the spectral range of 400–2000 cm^{-1} was selected for data presentation, with the results shown in Figure 12.

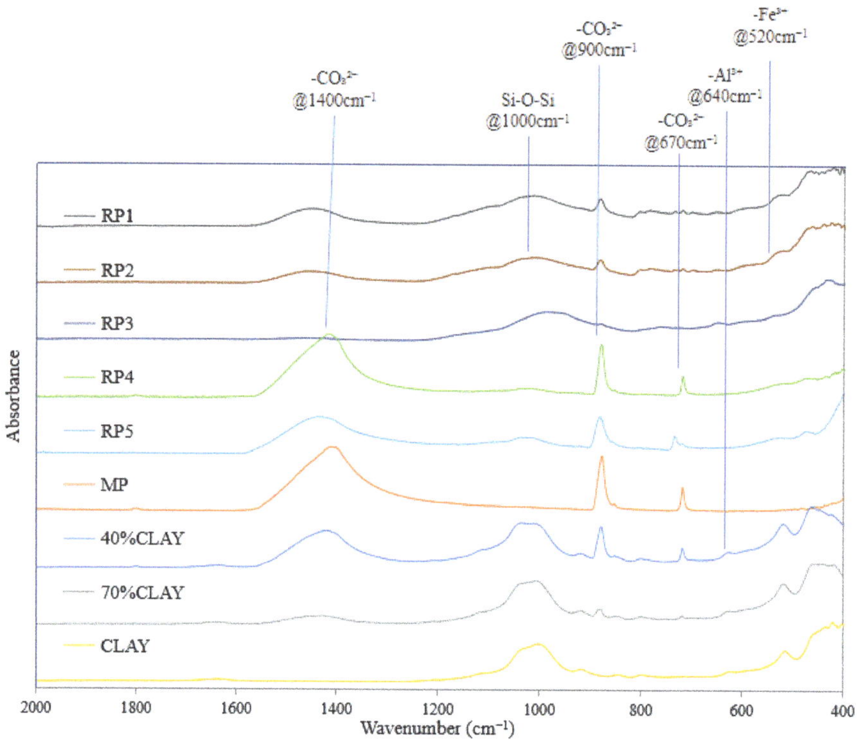

Figure 12. FTIR spectra of nine powders.

An analysis of Figure 12 with reference to related studies [25] identifies the characteristic functional group peaks as follows: the broad peak at 1000 cm^{-1} corresponds to the silicon–oxygen–silicon (Si-O-Si) stretching vibration, while Buzgar Nicolae et al. [26] confirmed that the peak at 1400 cm^{-1} represents the antisymmetric stretching vibration of carbonate groups (-CO_3^{2-}). Additional carbonate-related peaks include the out-of-plane bending vibration at 900 cm^{-1} and the in-plane bending vibration at 670 cm^{-1}. Furthermore, another study [27] assigned the peaks at 520 cm^{-1} and 640 cm^{-1} to trivalent iron (-Fe^{3+}) and aluminum ions (-Al^{3+}), respectively.

The analysis reveals a clear linear relationship between clay content and FTIR absorbance intensities for both carbonate (CO_3^{2-}) and siloxane (Si-O-Si) functional groups across the four reference samples (mineral powder, 40% clay, 70% clay, and pure clay). Specifically, RP4 and RP5 exhibit strong carbonate peaks (1400, 900, 670 cm^{-1}) with minimal Si-O-Si absorption (1000 cm^{-1}); RP3 shows pronounced peaks for both functional groups; and RP1 and RP2 demonstrate moderate absorbance for both peaks. These spectral patterns align with previous XRF, XRD, and MBV test results, confirming the consistency of our multi-technique characterization approach for evaluating reclaimed powder properties.

The experimental results demonstrate that the addition of clay simultaneously increases filler acidity and reduces cleanliness, necessitating strict control of clay content in reclaimed powder applications. Since aluminosilicates constitute the primary components of clay, a quantitative assessment of clay content can be achieved through characterizing the Si-O-Si absorption peak at 1000 cm^{-1} in the FTIR spectra.

The filler composition, predominantly comprising silicon and calcium components, allows for a quantitative assessment using characteristic carbonate absorption peaks observed at 1400 cm^{-1} (strong and broad antisymmetric stretching vibration), 900 cm^{-1} (weak out-of-plane bending vibration), and 670 cm^{-1} (weak in-plane bending vibration). The 1400 cm^{-1} peak proves particularly advantageous for quantitative analysis due to its high intensity, broad spectral window, and absence of overlapping interference from other functional groups. While the smaller 900 cm^{-1} and 670 cm^{-1} peaks are also indicative of carbonate content, they present challenges in differentiation and quantification due to their lower intensities and potential spectral overlaps. This spectral characterization approach enables precise carbonate content determination through 1400 cm^{-1} peak area integration.

Based on the aforementioned analysis, the antisymmetric stretching vibration peak of carbonate at 1400 cm^{-1} (hereinafter referred to as anti-carbonate) and the silicon–oxygen–silicon (Si-O-Si) absorption peak at 1000 cm^{-1} were selected from multiple functional group absorption peaks as the characteristic peaks in the infrared spectrum of reclaimed powder, serving as indicators for material acidity/alkalinity and cleanliness. Furthermore, in Fourier transform infrared spectroscopy, two common approaches are employed, namely a direct comparison of peak intensities and a semi-quantitative analysis using peak area ratios. Semi-quantitative analysis does not introduce new functional groups; it eliminates the influence of the sample amount on the spectral amplitude by the ratio [28]. The ratio of these two peak intensities (Si-O-Si to anti-carbonate and anti-carbonate to Si-O-Si) was adopted as an additional indicator for acidity/alkalinity. Since the area under the peak is a quantitative characterization of the functional group concentration standard, MATLAB programs were utilized to process the spectral data and automatically calculate functional group indices, with the processing workflow illustrated in Figure 13 and the calculation formulas provided in Equations (1)–(4). The results are presented in Table 4, where the zero value indicates concave spectral features in the corresponding regions, suggesting the absence of detectable absorption peaks. Subsequently, this study optimized the four quantitative indicators based on actual test results from XRF, pull-off tests, and methylene blue tests, ultimately proposing the evaluation indicator with the best predictive performance.

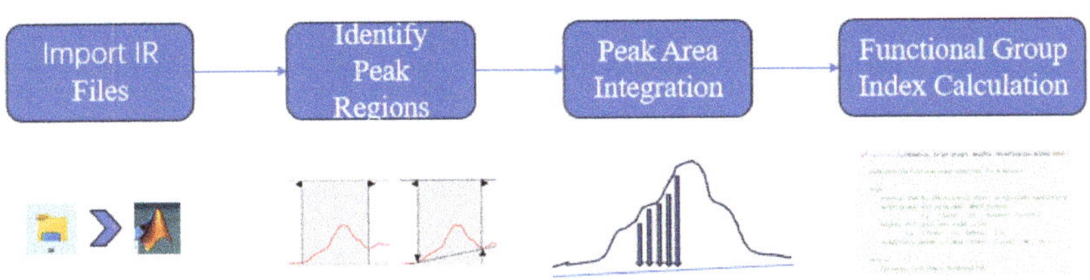

Figure 13. Schematic of spectral data processing steps.

$$I_C = A_{1400} \tag{1}$$

$$I_S = A_{1000} \tag{2}$$

$$I_{S/C} = \frac{A_{1000}}{A_{1400}} \tag{3}$$

$$I_{C/S} = \frac{A_{1400}}{A_{1000}} \tag{4}$$

Nomenclature:

I_C: Carbonate functional group index (1310 cm^{-1}~1550 cm^{-1});

I_S: Siloxane functional group index (930 cm^{-1}~1070 cm^{-1});

$I_{S/C}$: Siloxane-to-carbonate ratio index;

$I_{C/S}$: Carbonate-to-siloxane ratio index;
A_{XXX}: Integrated peak area at XXX cm^{-1} wavenumber.

Table 4. Functional group indices of powders extracted via MATLAB semi-quantitative analysis.

	I_C	I_S	$I_{S/C}$	$I_{C/S}$
RP1	2.35	5.40	2.30	0.44
RP2	1.14	3.18	2.79	0.36
RP3	0.68	14.01	20.60	0.05
RP4	19.93	1.77	0.09	11.26
RP5	10.85	2.85	0.26	3.81
MP	27.17	0	0	/
40% CLAY	17.4	21.1	1.21	0.82
70% CLAY	7.5	39.9	5.32	0.19
CLAY	0.0	61.9	/	0.0

3.5.2. Optimization of Evaluation Indicators for Acidity/Alkalinity (Based on XRF and Pull-Off Tests)

To verify the correlation between the acidity/alkalinity indicators extracted from the FTIR tests and the actual material properties, a correlation analysis was conducted between the calculated functional group indices and XRF results, with I_C and I_S analyzed in standard Cartesian coordinates, while $I_{S/C}$ and $I_{C/S}$ were evaluated in logarithmic coordinates. The correlation analysis mainly focuses on empirical formula fitting to establish a predictive relationship for the performance of the recycled powder [29], as shown in Figure 14.

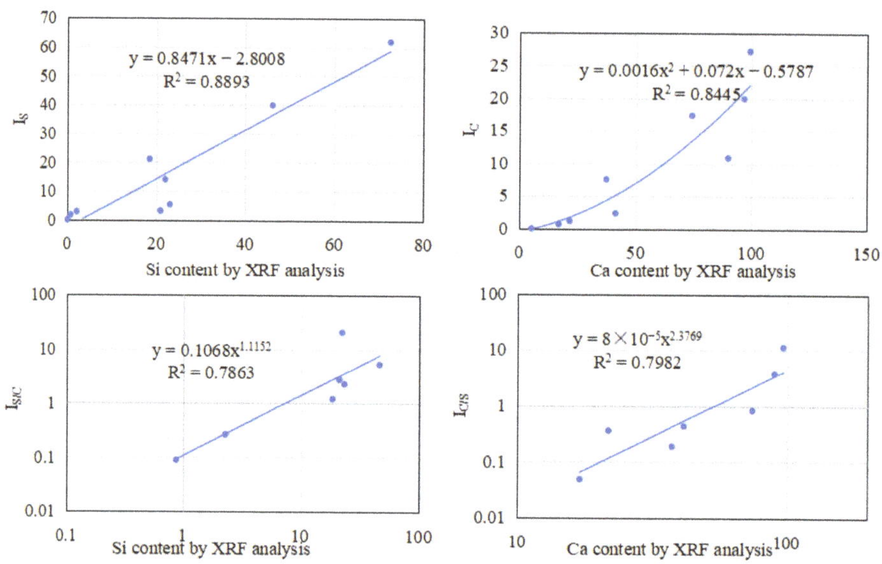

Figure 14. Correlation analysis between FTIR indices (I_S, I_C, $I_{S/C}$, and $I_{C/S}$) and corresponding XRF results.

Except for $I_{S/C}$, which showed a weaker correlation with XRF data, the other three indices (I_C, I_S, and $I_{C/S}$) all exhibited quite good correlations with the filler's actual acidity/alkalinity, particularly I_C and I_S, which achieved superior correlation coefficients (R^2 = 0.8893 and 0.8445, respectively). Therefore, they were selected for subsequent analysis.

The acidity/alkalinity of fillers significantly influences the adhesive properties of asphalt mastics. The reason for this phenomenon may be that the surface of clay minerals rich in hydroxyl (-OH) adsorbs water molecules and forms an interfacial water film with a thickness of 5–10 nm. This water film physically separates the asphalt from the aggregate, thereby reducing the adhesion of the asphalt [30].

Therefore, in addition to XRF testing, pull-off tests were conducted on asphalt mastics incorporating different reclaimed powders. ESSO 70# asphalt (produced by ExxonMobil, Houston, TX, USA) was heated to 150 °C and mixed with nine types of powder materials at a 25% filler content. The fluid asphalt mastic was then dropped onto a basalt aggregate substrate, covered with a pull-head, and subjected to controlled pressure. The pull-head design featured a 0.2 m protruding edge at the bottom to serve as a support, allowing for excessive asphalt mastic to flow out through overflow channels. The pull-head configuration for the pull-off test is illustrated in Figure 15.

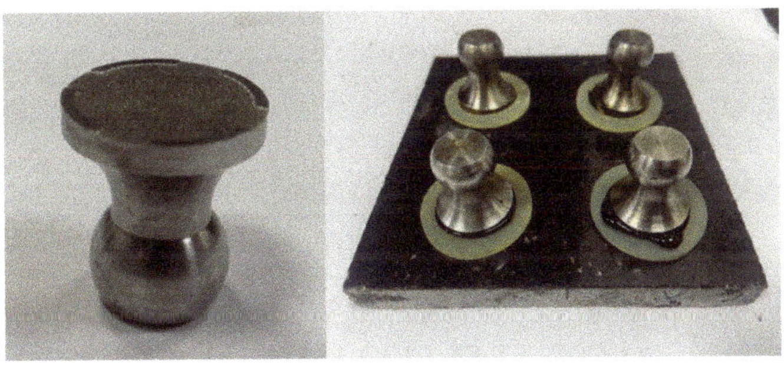

Figure 15. Pull-off test pull-head configuration and test specimen.

The specimens were then cured in a constant-temperature chamber at 25 °C for 2 h before being tested using a Positest AT-A pull-off adhesion tester, which applied vertical tension to the pull-heads bonded to the stone substrates. The pull-off strength, recorded after adhesive failure, served as an indicator of both the cohesive and adhesive properties of the asphalt, with each sample tested in triplicate and the results averaged. Subsequent correlation analysis between the FTIR-derived functional group indices (I_C and I_S) and the pull-off strength (Figure 16) revealed strong relationships (R^2 = 0.7608 for I_C and 0.803 for I_S), demonstrating that the acidity/alkalinity indicators obtained through infrared spectroscopy effectively correlate with the bonding performance of asphalt mastics. These findings confirm that FTIR spectroscopy can reliably predict both the chemical properties (acidity/alkalinity) and macroscopic mechanical behavior (adhesion) of fillers, providing a rapid assessment method for quality control in practical applications.

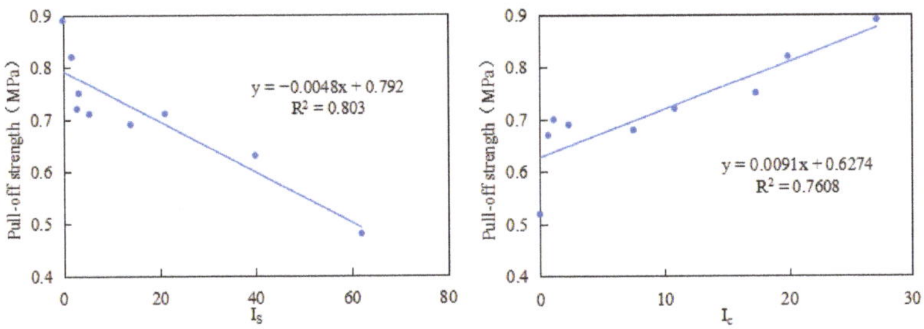

Figure 16. Correlation between FTIR index (I_S, I_C) and pull-off strength.

3.5.3. Optimization of Cleanliness Evaluation Indicators (Based on Methylene Blue Test)

The evaluation indicators for acidity/alkalinity demonstrated that both I_C and I_S exhibited good correlation with pH properties (R^2 > 0.84). This section further evaluates their effectiveness for cleanliness evaluation by conducting correlation analyses between I_C and I_S and MBV. The results are shown in Figure 17.

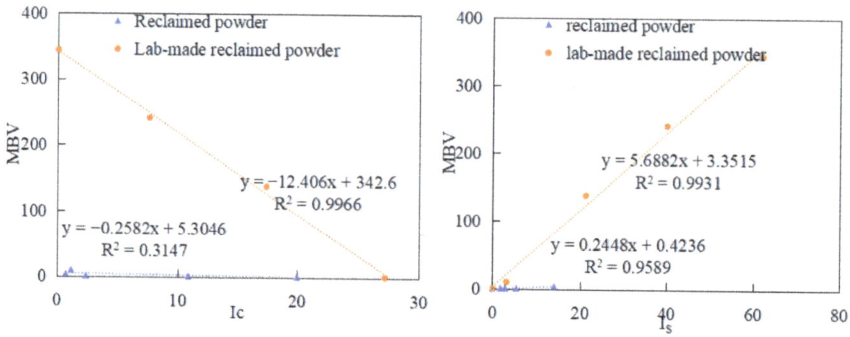

Figure 17. Correlation analysis of FTIR indices (I_S, I_C) and methylene blue values.

The analysis first reveals that the addition of clay significantly increases methylene blue values (MBVs) (orange data points in the figure), demonstrating its substantial impact on cleanliness. While the I_S index shows strong predictive capability for both clay-rich samples and conventional reclaimed powder samples, the I_C index performs well only for high-clay-content samples but performs poorly for standard reclaimed powders, indicating limited sensitivity. Given that field-collected reclaimed powders are more representative of real-world conditions, the I_S index is recommended as the preferred predictor for practical applications.

The results from the methylene blue tests, XRF analysis, and pull-off tests collectively demonstrate that the I_S index effectively predicts both the acidity/alkalinity and cleanliness of fillers, which can be attributed to the aluminosilicate composition of clay—while filler acidity/alkalinity is primarily influenced by silicon oxide (SiO_2) content and cleanliness is governed by clay content, both parameters fundamentally correlate with the silicon content in fillers, thereby making the Is index (quantifying Si-O-Si functional groups at 1000 cm^{-1}) a robust predictive indicator for these two critical properties, as evidenced by its strong correlations with SiO_2 content (R^2 = 0.87), MBV (R^2 = 0.79) and pull-off strength (R^2 = 0.81), confirming the dual diagnostic capability of this FTIR-derived spectral index for the simultaneous quality assessment of reclaimed powder materials.

4. Conclusions

To improve reclaimed powder utilization from asphalt plants, this study analyzed five field-collected and four artificial samples using SEM/XRF/XRD to characterize their properties. FTIR was then used to establish correlations between key parameters (acidity/alkalinity and cleanliness) and specific peak intensities, developing functional group indices validated by XRF, methylene blue, and pull-off tests. The main conclusions are as follows:

1. The reclaimed powder mainly consists of tiny, crushed stone particles and dust, with significant variations in crystal structure and chemical composition, including calcium carbonate, silicon oxide, iron oxide, aluminum oxide, etc. Some samples also contain clay, which is the primary factor affecting their properties. The presence of clay

increases the acidity of the filler while reducing its cleanliness. Since the main component of clay is aluminosilicate, the silicon–oxygen–silicon (Si-O-Si) characteristic peak at 1000 cm^{-1} can be used to quantitatively evaluate clay content.

2. In the infrared spectrum, the Si-O-Si absorption peak at 1000 cm^{-1} and the antisymmetric stretching peak of carbonate at 1400 cm^{-1} show excellent correlation with the calcium carbonate and silicon oxide content in reclaimed powder. These peaks can be used to predict the filler's calcium carbonate content, clay content, acidity/alkalinity, and cleanliness. Higher calcium carbonate and lower silicon oxide content indicate less clay, stronger alkalinity, and better cleanliness, making the material more suitable for reuse.
3. Based on the Si-O-Si peak at 1000 cm^{-1} and the carbonate antisymmetric stretching peak at 1400 cm^{-1}, four functional group indices (I_C, I_S, $I_{S/C}$, $I_{C/S}$) were calculated. Their predictive performance for acidity/alkalinity and cleanliness was compared using XRF, methylene blue tests, and pull-off tests, ultimately selecting the I_S index due to its strong correlation. The I_S index exhibited correlations of R^2 = 0.89 with XRF, R^2 = 0.80 with methylene blue values, and R^2 = 0.96 with pull-off strength, confirming its effectiveness in predicting both acidity/alkalinity and cleanliness.
4. Methylene blue tests also revealed that, in the absence of clay, dust and organic residues are the main factors influencing the MBV. However, even trace amounts of clay cause a sharp increase in the MBV, showing a completely different trend compared to clay-free samples. Therefore, during transportation and storage, reclaimed powder should be prevented from contact with clay.

Author Contributions: Conceptualization, C.Y.; Methodology, H.W., C.Y. and P.Z.; Software, R.X. and H.Z.; Validation, W.W.; Formal analysis, D.Y. and R.C.; Investigation, H.W. and D.Y.; Resources, R.X. and H.Z.; Data curation, R.X. and R.C.; Writing—original draft, D.Y.; Writing—review & editing, C.Y. and P.Z.; Visualization, P.Z.; Supervision, W.W., R.X. and R.C.; Project administration, H.W. and H.Z.; Funding acquisition, W.W. All authors have read and agreed to the published version of the manuscript.

Funding: The authors gratefully acknowledge the financial supports from the National key R&D program of China (2022YFB2602603), the National Natural Science Foundation of China (52478469), the Sichuan Youth Science and Technology Innovation Research Team (2022JDTD0015), and the Sichuan Transportation Science and Technology Program (2021-ZL-16, 2024-ZL-08).

Institutional Review Board Statement: Not applicable.

Informed Consent Statement: Not applicable.

Data Availability Statement: The original contributions presented in this study are included in the article. Further inquiries can be directed to the corresponding authors.

Conflicts of Interest: Authors Hao Wu and Wentao Wang were employed by Shandong Provincial Road and Bridge Group Co., Ltd. Authors Rui Xiao and Rong Chen were employed by Sichuan Transportation Construction Group Co., Ltd. Author Hengji Zhang was employed by Shanxi Transportation Holdings Science & Technology Transformation Co., Ltd. The remaining authors declare that the research was conducted in the absence of any commercial or financial relationships that could be construed as a potential conflict of interest.

References

1. Chaiyaput, S.; Sertsoongnern, P.; Ayawanna, J. Utilization of waste dust from asphalt concrete manufacturing as a sustainable subbase course material in pavement structures. *Sustainability* **2022**, *14*, 9804. [CrossRef]
2. Abelson, J. Powder Reclamation Methods. In *Modern Cold Spray: Materials, Process, and Applications*; Springer: Cham, Switzerland, 2015; pp. 317–340.

3. Sun, Y.; Huang, W.; Xu, G.; Gan, L.; Shen, Z.; Luo, S.; Zhang, H.; Feng, Y.; Xiong, H.; Shu, Y. Experimental investigation on the performance of hydraulic asphalt concrete with varying epoxy system content. *Constr. Build. Mater.* **2025**, *484*, 141755. [CrossRef]
4. Kockelman, K.M.; Siethoff, B.; Walton, M.; Mahmassani, H. *Research on Relationships Between Transportation Infrastructure & Increase in Vehicle Miles Traveled: The Effects of Highway Capacity Expansion on Land Development*; Final Technical Report for the US Environmental Protection Agency; Center for Transportation Research, University of Texas: Austin, TX, USA, 2002.
5. Ayawanna, J.; Suksawat, T.; Sertsoongnern, P.; Chaiyaput, S. The use of asphalt waste dust for stabilization of sustainable pavement recycling. *Constr. Build. Mater.* **2024**, *446*, 137936. [CrossRef]
6. Muniandy, R.; Aburkaba, E.E. Effects of filler type and particle size on permanent deformation of Stone Mastic Asphalt (SMA) mixtures. *Chem. Eng.* **2012**, *49*, 9902–9909.
7. Chen, C. *A Mineralogical Approach to Use the Non-Qualified Fine Aggregates in Asphalt Concrete Pavement*; Université Paris-Est: Créteil, France, 2016.
8. Wilanowicz, J.; Grabowski, W.; Andrzejczak, M.; Chromiec, A. Assessment of the suitability of baghouse dusts from a dust extractor as fillers for hot-mix asphalt. *Procedia Eng.* **2013**, *57*, 1269–1277. [CrossRef]
9. Mukhtar, N.; Hasan, M.R.M.; Shariff, K.A.; Bergh, W.V.D. Relationship between the physicochemical and electrostatic charge characteristics of filler materials on the morphological and adhesive Pull-Off tensile strength of asphalt mastics. *Constr. Build. Mater.* **2022**, *346*, 128343. [CrossRef]
10. Fowkes, F.M. Role of acid-base interfacial bonding in adhesion. *J. Adhes. Sci. Technol.* **1987**, *1*, 7–27. [CrossRef]
11. Xuan, W. *Study on the High Temperature Performance of Composite Modified Hard Asphalt Mixture*; Nanjing Forestry University: Nanjing, China, 2023.
12. Gajewski, M.D.; Horodecka, R.; Bańkowski, W.; Mirski, K.; Grzegórska, A.; Kłopociński, M. Effects of Impurities and Ageing on the Functional and Rheological Properties of Asphalts with Additives from Recovered and Pyrolysis-Processed Plastics. *Materials* **2024**, *17*, 3451. [CrossRef]
13. Biro, S.; Gandhi, T.; Amirkhanian, S. Determination of zero shear viscosity of warm asphalt binders. *Constr. Build. Mater.* **2009**, *23*, 2080–2086. [CrossRef]
14. Topçu, I.B.; Demir, A. Relationship between methylene blue values of concrete aggregate fines and some concrete properties. *Can. J. Civ. Eng.* **2008**, *35*, 379–383. [CrossRef]
15. Hou, X.; He, Y.; Jones, B.T. Recent advances in portable X-ray fluorescence spectrometry. *Appl. Spectrosc. Rev.* **2004**, *39*, 1–25. [CrossRef]
16. Moreno-Maroto, J.M.; Alonso-Azcárate, J.; O'Kelly, B.C. Review and critical examination of fine-grained soil classification systems based on plasticity. *Appl. Clay Sci.* **2021**, *200*, 105955. [CrossRef]
17. Singh, S.K.; Kathait, N.; Kirthika, S.K. Clay Content Assessment in Fine Aggregates Using Phenol Adsorption Method. *J. Mater. Civ. Eng.* **2022**, *34*, 04022044. [CrossRef]
18. Luo, S.; Tian, J.; Liu, Z.; Lu, Q.; Zhong, K.; Yang, X. Rapid determination of styrene-butadiene-styrene (SBS) content in modified asphalt based on Fourier transform infrared (FTIR) spectrometer and linear regression analysis. *Measurement* **2020**, *151*, 107204. [CrossRef]
19. Kępniak, M.; Woyciechowski, P.; Franus, W. Chemical and physical properties of limestone powder as a potential microfiller of polymer composites. *Arch. Civ. Eng.* **2017**, *63*, 67–78. [CrossRef]
20. Zhou, S.; Long, K.; Zhang, Z.; Li, S.; Ai, C.; Yan, C. Development of sustainable Lignin-Based coatings for layered double Hydroxides: Enhancing synergistic Anti-Aging properties in bitumen. *Fuel* **2025**, *380*, 133166. [CrossRef]
21. *JTG 3432-2024*; Ministry of Transport of the People's Republic of China. Test Methods of Aggregate for Highway Engineering. China Communications Press: Beijing, China, 2024.
22. Bergna, H.E. Colloid chemistry of silica: An overview. *Adv. Chem.* **1994**, *234*, 1–47.
23. Wang, L.; Zhang, L.; Liu, Y. Molecular Dynamics Study on the Effect of Mineral Composition on the Interface Interaction between Rubberized Asphalt and Aggregate. *J. Mater. Civ. Eng.* **2022**, *34*, 04022032. [CrossRef]
24. Yin, Y.; Chen, H.; Kuang, D.; Song, L.; Wang, L. Effect of chemical composition of aggregate on interfacial adhesion property between aggregate and asphalt. *Constr. Build. Mater.* **2017**, *146*, 146231–146237. [CrossRef]
25. Socrates, G. *Infrared and Raman Characteristic Group Frequencies: Tables and Charts*; John Wiley & Sons: Hoboken, NJ, USA, 2004.
26. Buzgar, N.; Apopei, A.I. The Raman study of certain carbonates. *Geologie* **2009**, *LV*, 97–112.
27. Biswas, B.; Allen, H.C. Solution and Surface Solvation of Nitrate Anions with Iron (III) and Aluminum (III) in Aqueous Environments: A Raman and Vibrational Sum Frequency Generation Study. *J. Phys. Chem. A* **2024**, *128*, 8938–8953. [CrossRef]
28. Ngezahayo, E.; Torbaghan, M.E.; Metje, N.; Burrow, M.; Ghataora, G.S.; Desalegn, Y. Investigating the Effectiveness of Fourier Transform Infrared Spectroscopy (FTIR) as an Antifraud Approach for Modified Epoxy Asphalt Mixes in Developing Countries. *Sustainability* **2023**, *15*, 16332. [CrossRef]

29. Weigel, S.; Stephan, D. The prediction of bitumen properties based on FTIR and multivariate analysis methods. *Fuel* **2017**, *208*, 655–661. [CrossRef]
30. Zhang, J.; Apeagyei, A.K.; Airey, G.D.; Grenfell, J.R. Influence of aggregate mineralogical composition on water resistance of aggregate–bitumen adhesion. *Int. J. Adhes. Adhes.* **2015**, *62*, 45–54. [CrossRef]

materials

MDPI

Article

Evaluation and Control of Variability in RAP Properties Through Refined Fractionation Processing Methods

Yan Zhang [1], Jiyang Li [2] and Yiren Sun [2,*]

[1] City Institute, Dalian University of Technology, Dalian 116600, China
[2] School of Infrastructure Engineering, Dalian University of Technology, Dalian 116024, China
* Correspondence: sunyiren@dlut.edu.cn

Abstract

Variability in reclaimed asphalt pavement (RAP) properties, such as aggregate gradation, asphalt content, and moisture content, poses a significant challenge to producing consistent and reliable recycled asphalt mixtures. This study systematically evaluated processing techniques for mitigating variability through a comparative analysis of four fractionation strategies, i.e., unfractionated, two-fraction, four-fraction, and six-fraction processing. Corresponding to the four approaches, four distinct reference RAP mixtures were fabricated by proportionally recombining the obtained RAP fractions towards a target gradation. The gray relational analysis (GRA) was employed to quantify geometric similarity between the gradation curve of reclaimed aggregates from each fraction and the target gradation curve, thereby facilitating efficient determination of blending proportions without resorting to complex optimization algorithms. Statistical variability indicators, including range, standard deviation, and coefficient of variation (COV), were used to assess the effectiveness of each fractionation and recombining method. The results demonstrated that refined fractionation processing significantly reduced variability, particularly in gradation properties. Compared with the COV values from the commonly used two-fraction processing, those from the refined four-fraction and six-fraction processing methods decreased by up to 51.5% and 73.5%, respectively. While increasing the number of fractions generally enhanced homogeneity, the four-fraction approach emerged as the most technically reliable and economically viable strategy, achieving a desirable balance between processing effort and variability control. Furthermore, the GRA proved to be a practical and efficient tool for blend proportioning, reducing reliance on complex numerical methods. These findings reveal the importance of refined fractionated RAP processing in enabling the production of high-RAP recycled mixtures with improved uniformity and performance.

Keywords: reclaimed asphalt pavement; variability; fractionation process; gradation; asphalt concrete

1. Introduction

The utilization of reclaimed asphalt pavement (RAP) in new asphalt mixtures is growing increasingly widespread within the pavement industry [1–3]. The RAP material, obtained by milling or removing deteriorated pavements, consists of valuable crushed aggregates and hardened aged binder. Its reuse supports natural resource conservation and cost reduction, thus offering considerable economic and environmental advantages. In recent years, the pursuit of sustainable pavement infrastructure has substantially increased the motivation to use high percentages of RAP in recycled mixtures [4–6]; however, one major concern is variability in RAP properties like aggregate gradation, asphalt content, and moisture content, which poses a severe challenge to the quality control of high-RAP recycled mixtures [7–9]. Montanez et al. [10] reported that variability in RAP materials prominently affects the moisture susceptibility of recycled mixes. Austerman et al. [11] further demonstrated that even low RAP contents (e.g., <15%) can lead to significant performance deviations if the variability of RAP is not properly controlled. As a consequence, many highway agencies restrict the RAP content in surface layers to levels below 25% to lower these potential risks [12,13].

Reducing the variability in RAP is critical to ensuring the performance of recycled mixtures, especially those with high RAP contents, since it not only complicates mix design but also adversely influences the homogeneity of mixtures [11,14,15]. There exist numerous factors that lead to RAP variability [16]. Typically, RAP materials are sourced from extensive pavement sections within a region, where variations in original material specifications, construction dates, traffic history, and environmental exposure create inherent heterogeneity. Furthermore, additional variability can be introduced during the milling and collection process. Finally, stockpiling, transportation, and handling operations can induce segregation and further alter material properties, compounding the overall variability [17,18]. From a mesoscopic viewpoint, both the gradation of old aggregates and the content and properties of aged binder in RAP vary greatly, and the RAP particle agglomeration further intensify the uncertainty of particle gradations and inconsistency of activated binder contents in recycling process [19–23].

To address the variability issue, researchers have explored various management and processing approaches for RAP materials [24,25]. Currently, appropriate management strategies primarily involve milling in layers to segregate diverse mix types, separating RAP into multiple stockpiles for different sources, and maintaining homogeneity by mixing RAP in stockpiles [18]. Although these practices are valuable, they mostly fail to capture the quantitative variation details on RAP consistency throughout homogenization operations. By contrast, processing methods, primarily through crushing (using impact crushers, jaw crushers, etc.) and fractionation (with various sieve sequences), are more straightforward strategies for reducing variability. These processing procedures are able to effectively break down agglomerations and control maximum particle size, thereby transforming variable RAP into a consistent recycled material. In addition, emerging techniques such as refined decomposition of aged binder and old aggregates in RAP offer future alternatives [26,27].

Among the aforementioned approaches, the fractionation processing for RAP materials, which divides the collected RAP into two or more stockpiles by particle size, appears to be the most commonly used one [28,29]. Research efforts have been made to evaluate and control the variability of RAP by various fractionation processing methods. Shen et al. [30] employed the 4.75 mm sieve as a critical sieve for two-fraction processing and the 4.75 mm and 2.36 mm sieves for the three-fraction processing. They reported that in terms of the Superpave volumetric requirements, incorporating an additional 2.36 mm sieve effectively improved the possible content of RAP by about 30%. To alleviate the negative effect of excessive fine aggregates in RAP, Shannon et al. [28] designed two dual-fraction (coarse and fine fractions) processing methods, and found that owing to the improved aggregate structure, the optimum asphalt content was lowered when the fractionation processing was involved. Katla et al. [4] analyzed the influence of different fractionating practices, including unfractionation, two-level fractionation, and three-level fractionation, and showed that three-level fractionation processing resulted in much lower variability in the volumetric properties and performance of recycled mixtures than those from the unfractionation and two-level fractionation methods.

Although the benefits of fractionation processing in reducing the RAP variability has been widely accepted, existing studies and practices mostly concentrated on the two-fraction method and detailed and quantitative investigations into the efficacy of refined fractionation processing operations remain very limited [4,28,29]. Thus, there is an urgent need to develop new methods that can appropriately evaluate and control RAP variability. To this end, this study presented an approach to effectively evaluate the effect of refined fractionation processing operations on the variability in properties of RAP materials as well as a recombining method by means of gray relational analysis (GRA) to accurately determine the blending proportions of all RAP fractions so that desired properties can be well achieved.

2. Materials

The RAP material utilized in this study was collected from the binder layer (middle surface layer) of an expressway located in Dalian, China. The material was obtained via cold milling of the asphalt pavement, which was designed as an AC-16 mixture with a nominal maximum aggregate size (NMAS) of 16 mm. The original binder was an unmodified asphalt with a penetration grade of 80/100, and the original coarse aggregates were basalt. The moisture content of the RAP material was measured as 1.27% in accordance with the test methods of aggregates for highway engineering of China [31]. In this study, the recycled asphalt mixture was considered as AC-16 with a NMAS of 16 mm.

The aged asphalt binder was extracted from the RAP material using trichloroethylene solvent, and recovered from the solution via a rotary evaporator [32]. In this manner, the asphalt content in the RAP was determined as 4.02% and the old RAP aggregates were obtained. The basic properties of the aged asphalt and old coarse aggregates in the RAP material are presented in Table 1. As observed, all the measurements of the properties satisfy the requirements for recycling [33]. Sieve analysis was conducted on both the RAP material and the reclaimed RAP aggregates. The resulting gradations are shown in Figure 1. As can be seen, the size distribution of the coarse aggregates in the RAP experienced significant variation due to long-term traffic loading and the mechanical action during milling.

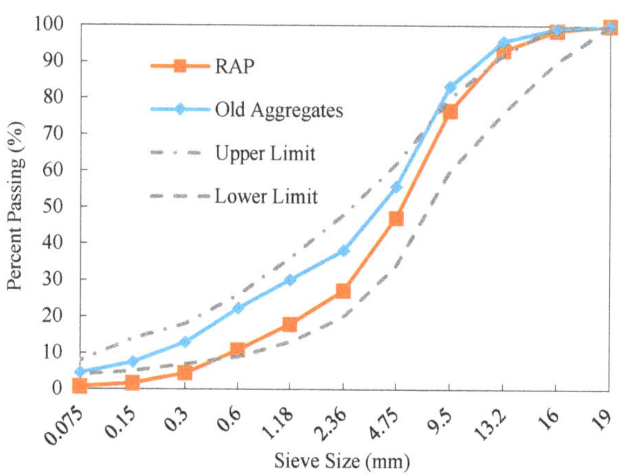

Figure 1. Gradations of the RAP material and extracted old aggregates.

Table 1. Properties of the aged asphalt and old coarse aggregates in the RAP material.

Material	Property	Requirement [33]	Measurement
Aged asphalt	Penetration at 25 °C (0.1 mm)	>20	27.9
	Softening point (°C)	—	67.1
	Ductility at 15 °C (mm)	—	89.8
	Apparent viscosity at 135 °C (Pa·s)	—	0.697
Old coarse aggregate	Flat and elongated particle fraction (%)	≤15	14.1
	Aggregate crushing value (%)	≤26	14.4

3. Methodology

3.1. Fractionation Processing Methods

Fractionation is one of the most commonly used processing approaches for mitigating the variability in RAP properties. In current engineering practice, RAP is generally processed into two fractions, i.e., the coarse and fine ones; however, there would be still considerable heterogeneity in recycled mixtures, especially for the utilization of high percentages of RAP. For this reason, refined fractionation methods were adopted to achieve better control of the RAP variability.

In the present study, four scenarios were taken into account to investigate the effect of various fractionation processing methods. In Scenario 1, the RAP was unfractionated; that is, the particle size of the RAP ranged from 0 to 19 mm. In Scenario 2, the RAP material was separated into two stockpiles by the sieve of 4.75 mm as the common practice, resulting in two fractions, 0~4.75 mm and 4.75~19 mm. The sieve of 4.75 mm was chosen as a critical sieve mainly due to a common practice for aggregate fractionation processing in China. In Scenario 3, the fractionation process for the RAP was refined by four stockpiles, namely, 0~4.75 mm, 4.75~9.5 mm, 9.5~13.2 mm, and 13.2~19 mm. In Scenario 4, six detailed fractions were considered, i.e., 0~2.36 mm, 2.36~4.75 mm, 4.75~9.5 mm, 9.5~13.2 mm, 13.2~16 mm, and 16~19 mm. After fractionation processing, the fractionated RAP materials were recombined towards a target gradation. In this study, the average gradation (median) between the upper and lower limits for AC-16 was taken as the target gradation for the blended aggregates of the recycled mixture.

3.2. Proportioning RAP Fractions into Reference Materials

Corresponding to the four fractionation processing methods mentioned above, four different reference RAP materials, denoted as Reference-1, Reference-2, Reference-3, and Reference-4, were obtained by proportioning the RAP factions in terms of the target gradation, as listed in Table 2. The desired gradation can be attained by combining the old aggregates recovered from the reference RAP materials and the virgin aggregates that have been divided into sieve sizes. In this way, the variability in RAP can be well controlled, and the effects of different fractionation methods can be revealed in detail.

Table 2. Reference RAP materials from different fractionation processing methods.

Reference RAP	Number of Fractions	Particle Sizes for RAP Fractions (mm)
Reference-1	1	0~19
Reference-2	2	0~4.75 and 4.75~19
Reference-3	4	0~4.75, 4.75~9.5, 9.5~13.2 and 13.2~19
Reference-4	6	0~2.36, 2.36~4.75, 4.75~9.5, 9.5~13.2, 13.2~16 and 16~19

To obtain the optimal proportions of the RAP fractions, the GRA [34,35] was employed in this study. The GRA, which was proposed by Professor Julong Deng, offers notable advantages, including flexibility with sample size, minimal distribution requirements, low computational cost, and consistent integration of quantitative and qualitative results. The underlying principle of the GRA is that the strength of the relationship between data sequences is governed by the similarity of their geometric curves. Greater geometric similarity corresponds to a stronger relational grade. As such, the RAP fraction with greater similarity in the gradation curve of old aggregates to the target gradation curve can be considered to account for a larger proportion in the reference RAP material. Based on this theory, the proportions of all fractions of old aggregates can be rationally and readily determined. Detailed information about GRA can be found elsewhere [34].

Since Reference-1 was an unfractionated material, it was taken as a control material and the GRA was applied only to References-2 through 4. Before performing the GRA, the old aggregates in all the RAP fractions from different fractionation processing methods were recovered by the ignition method [36]. The specific computational steps are described as follows:

(1) Let $Y(i)$ be the percent passing at each sieve size for the target gradation, and $X_j(i)$ be the percent passing of the j-th fractionated RAP material at the corresponding sieve size, where i denotes the sieve number ($i = 1, 2, \ldots, n$) and j represents the number of the RAP fractions ($j = 1, 2, \ldots, k$). For example, the gradations of the old aggregates from the six-fraction processing method are shown in Table 3. Since the data sequences shared the same dimensionality, the conventional normalization process was omitted in this study.

Table 3. Target gradation and gradations of old aggregates from the six-fraction processing method.

Sieve Size (mm)	Target (%)	Fraction 1 (%)	Fraction 2 (%)	Fraction 3 (%)	Fraction 4 (%)	Fraction 5 (%)	Fraction 6 (%)
19	100.0	100.0	100.0	100.0	100.0	100.0	100.0
16	95.0	100.0	100.0	100.0	100.0	100.0	47.4
13.2	84.0	100.0	100.0	100.0	100.0	43.7	37.7
9.5	70.0	100.0	100.0	100.0	29.2	27.2	30.6
4.75	48.0	100.0	100.0	25.5	16.9	23.3	22.0
2.36	34.0	100.0	22.4	15.3	13.2	20.2	15.9
1.18	24.5	71.2	15.9	13.7	11.3	16.3	11.8
0.6	17.5	46.5	13.5	11.6	9.3	12.8	8.3
0.3	12.5	21.8	9.3	7.7	6.0	7.8	4.4
0.15	9.5	11.4	6.3	4.9	3.7	4.8	2.3
0.075	6.0	7.0	4.8	3.4	2.5	3.3	1.3

(2) Calculate the difference sequences $d_j(i)$ between $Y(i)$ and $X_j(i)$, as follows:

$$d_j(i) = |Y(i) - X_j(i)| \tag{1}$$

For instance, the difference sequences for the Reference-4 RAP blend is presented in Table 4.

Table 4. Difference sequences between the target gradation and the gradations of old aggregates from the six-fraction processing method.

Sieve Size (mm)	Fraction 1 (%)	Fraction 2 (%)	Fraction 3 (%)	Fraction 4 (%)	Fraction 5 (%)	Fraction 6 (%)
16	5.0	5.0	5.0	5.0	5.0	47.6
13.2	16.0	16.0	16.0	16.0	40.3	46.3
9.5	30.0	30.0	30.0	40.8	42.8	39.4
4.75	52.0	52.0	22.5	31.1	24.7	26.0

Table 4. Cont.

Sieve Size (mm)	Fraction 1 (%)	Fraction 2 (%)	Fraction 3 (%)	Fraction 4 (%)	Fraction 5 (%)	Fraction 6 (%)
2.36	66.0	11.6	18.7	20.8	13.8	18.1
1.18	46.7	8.6	10.8	13.2	8.2	12.7
0.6	29.0	4.0	5.9	8.2	4.7	9.2
0.3	9.3	3.2	4.8	6.5	4.7	8.1
0.15	1.9	3.2	4.6	5.8	4.7	7.2
0.075	1.0	1.2	2.6	3.5	2.7	4.7

(3) Determine the maximum value ($d_{\max}$) and minimum value ($d_{\min}$) of the difference sequences:

$$d_{\max} = \max_j \max_i d_j(i) \tag{2}$$

$$d_{\min} = \min_j \min_i d_j(i) \tag{3}$$

The calculated values are shown in Table 5.

Table 5. Maximum and minimum values of the difference sequences of the reference RAP blends.

Material	$d_{\max}$ (%)	$d_{\min}$ (%)
Reference-2	52.0	0.3
Reference-3	52.0	0.1
Reference-4	66.0	1.0

(4) Calculate the gray relational coefficient $\lambda_j(i)$, gray relational degree λ_j, and proportion of each fraction of the old aggregates ρ_j using the following equation:

$$\lambda_j(i) = \frac{d_{\min} + \xi d_{\max}}{d_j(i) + \xi d_{\max}} \tag{4}$$

$$\lambda_j = \frac{\sum_{i=1}^{n} \lambda_j(i)}{n} \tag{5}$$

$$\rho_j = \frac{\lambda_j}{\sum_{j=1}^{k} \lambda_j} \tag{6}$$

where $\xi \in (0,1)$ is the distinguishing coefficient, and in this study ξ is assumed to be 0.3. Table 6 displays the calculation results for the Reference-4 RAP material. It can be seen that the gradation of Fraction 2 has the greatest relational degree with the target gradation, followed by those of Fractions 3 and 5.

Table 6. Gray relational coefficients, gray relational degrees, and proportions of all fractions of old aggregates for Reference-4.

Quantity	Sieve Size (mm)	Fraction 1 (%)	Fraction 2 (%)	Fraction 3 (%)	Fraction 4 (%)	Fraction 5 (%)	Fraction 6 (%)
$\lambda_j(1)$	16	82.0	82.0	82.0	82.0	82.0	28.1
$\lambda_j(2)$	13.2	54.9	54.9	54.9	54.9	31.7	28.7
$\lambda_j(3)$	9.5	38.6	38.6	38.6	31.4	30.4	32.2
$\lambda_j(4)$	4.75	26.3	26.3	45.9	37.7	43.5	42.2
$\lambda_j(5)$	2.36	21.9	63.3	50.7	48.0	58.7	51.6

Table 6. *Cont.*

Quantity	Sieve Size (mm)	Fraction 1 (%)	Fraction 2 (%)	Fraction 3 (%)	Fraction 4 (%)	Fraction 5 (%)	Fraction 6 (%)
$\lambda_j(6)$	1.18	28.5	70.6	65.0	59.9	71.6	60.9
$\lambda_j(7)$	0.6	39.4	85.8	78.9	71.7	83.2	69.0
$\lambda_j(8)$	0.3	68.6	89.0	82.9	76.7	83.1	71.9
$\lambda_j(9)$	0.15	95.4	89.4	83.5	79.1	83.2	74.7
$\lambda_j(10)$	0.075	99.7	99.0	92.1	88.0	91.5	83.2
λ_j	—	55.5	69.9	67.4	62.9	65.9	54.2
ρ_j	—	14.8	18.6	17.9	16.7	17.5	14.4

(5) Let $Z(i)$ and $T(i)$ be the percent passing of old aggregates of a reference RAP blend and virgin aggregates at the i-th sieve. Calculate the gradations of the reference RAP blends and the corresponding ones of the virgin aggregates at a RAP content M using the following equations:

$$Z(i) = \sum_{j=1}^{k} [X_j(i) \times \rho_j] \tag{7}$$

$$T(i) = \frac{Y(i) - M \times Z(i)}{1 - M} \tag{8}$$

Figure 2 shows the calculated gradations of reclaimed aggregates of the four reference RAP materials. It can be observed that as the number of fractions increased, the gradation curve of old aggregate blends approached closer to the target gradation curve. The old aggregate gradation curves of References-3 and 4 were nearly identical, which indicates that from the perspective of aggregate blending, the processing methods that fractionate the RAP material into four and six fractions yielded comparable effects. Therefore, dividing RAP material into four fractions is more economically viable based solely on the desired gradation. It can be found that the GRA offers a simple and efficient approach for directly determining the proportion of each RAP fraction, avoiding the dependence on the nonlinear optimization algorithm for minimizing a target error function.

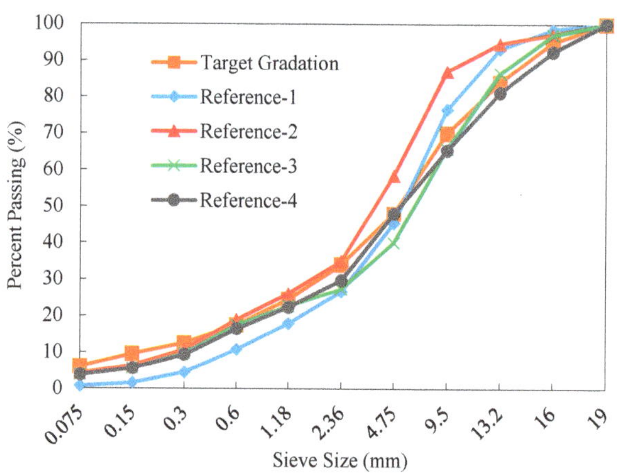

Figure 2. Calculated gradations of old aggregates of the four reference RAP materials.

3.3. Determining Moisture Contents of RAP Materials

RAP materials contain a certain amount of moisture. The presence of moisture not only prolongs the heating time of RAP, leading to further aging of the old asphalt, but also increases energy consumption during construction, thereby raising production costs. To ensure effective heating of RAP materials and improve production efficiency, the moisture content of RAP, w, is required to be less than 3% according to the technical specifications for highway asphalt pavement recycling of China [33], in which w is defined by the following:

$$w = \frac{m_w - m_d}{m_d} \times 100 \tag{9}$$

where m_w is the original mass of RAP (around 2000 g), and m_d is the constant mass of RAP after being dried in an oven at 105 °C.

3.4. Determining Asphalt Contents of RAP Materials

The asphalt contents of the RAP materials were determined using the ignition method [36]. A sample of RAP (about 2000 g) is weighed and placed in a furnace at 105 °C until it is dried to a constant mass. The mass of the dried RAP is recorded as m_1. The dried sample is then transferred to a preheated ignition furnace, where it is subjected to high temperatures between 482 °C and 538 °C until a constant mass m_2 is achieved. The mass loss is recorded as $m_1 - m_2$. The asphalt content of the RAP, Pb, is finally calculated by the following:

$$Pb = \frac{m_1 - m_2}{m_1} \times 100 \tag{10}$$

3.5. Evaluating RAP Variability

The coefficient of variation (COV) was adopted as the indicator for evaluating the RAP variability. The value of COV is jointly influenced by both the standard deviation S and the mean $\overline{x}$. Therefore, when using the COV to assess the uniformity of a dataset, both the mean $\overline{x}$ and the standard deviation S should be reported together. The COV is expressed by the following:

$$COV = \frac{S}{\overline{x}} \times 100 \tag{11}$$

$$\overline{x} = \frac{\sum_{\alpha=1}^{N} x_\alpha}{N} \tag{12}$$

$$S = \sqrt{\frac{\sum_{\alpha=1}^{N} (x_\alpha - \overline{x})^2}{N-1}} \tag{13}$$

where N is the number of sampling points. In addition, the range, Ra, of dataset x_α is defined by the following:

$$Ra = \max x_\alpha - \min x_\alpha \tag{14}$$

4. Results and Discussion

4.1. Variability Analysis of Moisture Contents of Reference RAP Materials

According to the proportions calculated form the GRA, the four reference RAP materials were produced. For each reference RAP material, moisture content tests were conducted using five 2000 g replicate samples. The moisture content test results and the calculated variability indicators of the four reference materials are presented in Tables 7 and 8. It can be observed that, compared to the unfractionated Reference-1 material, References-2, 3, and 4 exhibited significant reductions in the S, COV, and Ra of moisture content. In particular, compared to the COV value from the commonly used two-fraction processing, those from the four-fraction and six-fraction methods decreased by 39.8% and 47.6%, respectively. This indicates that the refined fractionation processing method can effectively reduce the variability in the moisture contents of RAP materials. As the number of fractions increases, the variability in the moisture content of the reference materials decreases accordingly; however, the extent of reduction diminishes with further increases in the number of fractions.

Table 7. Test results of moisture contents of reference RAP materials subjected to different fractionation processing methods.

Reference RAP	Replicate	m_w (g)	m_d (g)	w (%)
1	1	2000.0	1983.0	0.86
	2	2000.0	1982.1	0.90
	3	2000.0	1981.6	0.93
	4	2000.0	1978.3	1.10
	5	2000.0	1976.5	1.19
2	1	2000.0	1985.5	0.73
	2	2000.9	1987.3	0.68
	3	2001.3	1986.9	0.72
	4	2001.2	1988.3	0.65
	5	2001.2	1989.3	0.60
3	1	2001.7	1996.1	0.28
	2	2000.8	1995.0	0.29
	3	2000.7	1995.6	0.26
	4	2000.9	1995.6	0.27
	5	2000.0	1995.1	0.25
4	1	2001.2	1996.1	0.26
	2	2003.0	1997.6	0.27
	3	2003.2	1998.0	0.26
	4	2004.1	1998.2	0.30
	5	2001.5	1996.1	0.27

Table 8. Variation indicators of moisture contents of reference RAP materials.

Material	S (%)	$\bar{x}$ (%)	COV (%)	Ra (%)
Reference-1	0.47	4.47	10.46	1.07
Reference-2	0.16	4.58	3.49	0.44
Reference-3	0.09	4.09	2.10	0.21
Reference-4	0.07	4.00	1.83	0.18

4.2. Variability Analysis of Asphalt Contents of Reference RAP Materials

Asphalt content testing was carried out using five 2000 g replicate samples for each of the four reference RAP materials. Tables 9 and 10 show the asphalt content measurements and the calculated variability indicators of the four reference materials. It can be observed that, compared to Reference-1, References-2, 3, and 4 displayed a gradual decrease in the S, COV, and Ra of asphalt content. It is worth noting that compared with the COV value from the commonly used two-fraction processing, those from the four-fraction and six-fraction methods declined by 39.5% and 51.5%, respectively. This indicates that the refined fractionation processing approach not only effectively mitigates the variability in the asphalt contents of RAP materials, but demonstrates that a greater number of fractions results in improved homogeneity of the RAP material as well, which is consistent with findings from other researchers [4,28,29].

Table 9. Test results of asphalt contents of reference RAP materials subjected to different fractionation processing methods.

Reference RAP	Replicate	m_1 (g)	m_1–m_2 (g)	*Pb* (%)
1	1	1905.9	76.4	3.85
	2	1887.2	88.5	4.48
	3	1899.7	74.2	3.76
	4	1879.7	94.3	4.78
	5	1881.6	88.0	4.47
2	1	1887.0	87.5	4.43
	2	1892.5	87.3	4.41
	3	1892.0	85.7	4.33
	4	1899.5	82.7	4.17
	5	1893.0	90.7	4.57
3	1	1914.4	78	3.91
	2	1914.5	79.7	4.00
	3	1909.7	79	3.97
	4	1907.2	78.8	3.97
	5	1919.6	75.8	3.80
4	1	1919.7	75	3.76
	2	1916.3	78.1	3.92
	3	1918.8	76	3.81
	4	1918.1	77.5	3.88
	5	1916.1	77.1	3.87

Table 10. Variation indicators of asphalt contents of reference RAP materials.

Material	*S* (%)	$\overline{x}$	*COV* (%)	*Ra* (%)
Reference-1	0.44	4.27	10.31	1.02
Reference-2	0.15	4.38	3.34	0.40
Reference-3	0.08	3.93	2.02	0.20
Reference-4	0.06	3.85	1.62	0.16

4.3. Variability Analysis of Gradations of Reference RAP Materials and Old Aggregates

The old aggregates within RAP play an essential load bearing and transfer role along with virgin aggregates in recycled asphalt mixture. The gradation of the reclaimed aggregates directly impacts the stability of recycled mixture as a structural layer in pavement systems. If the gradation variability of the RAP material is high, it becomes difficult to ensure the consistency of the recycled mixture's gradation, which is one of the critical factors limiting the percentage of RAP in recycled asphalt mixtures. Therefore, it is crucial to experimentally evaluate the degree of variability in the gradation of RAP materials, as this serves as an important indicator for determining whether the RAP can be successfully utilized in recycled mixtures. In this section, variability assessments are conducted separately on the gradations of the four reference RAP materials and the gradations of the old aggregates within them.

Sieve analysis was performed using five 2000 g replicate samples for each of the four reference RAP materials. The specific test results of percentage retained of the RAP particles are presented in Table 11, and Figure 3 gives the range *Ra* (i.e., the difference between the maximum and minimum) of the percentage retained for each of the four reference materials. As observed, the range values of all percentage retained values for References-2, 3, and 4 met the requirement, less than 10%, specified in recycling technical specifications. In contrast, the range values over the sieve sizes of 1.18~9.5 mm for the unfractionated RAP material (Reference-1) exceeded 10%. This suggests that dividing RAP into just two fractions can reduce the dispersion of RAP gradation to a certain extent. In addition, it can be observed from the results for References-3 and 4 that the refined fractionation methods, i.e., the four-fraction and six-fraction approaches, further significantly alleviated the gradation divergence of the RAP particles smaller than the NMAS (13.2 mm). The range values for RAP materials across all particle sizes generally decreased as the number of fractions increased, although the extent of reduction gradually diminished. This implies that a refined fractionation yields lower variability in RAP materials, as well as improved homogeneity of the reference materials.

Table 11. Test results of percentage retained of the reference RAP materials subjected to different fractionation processing methods.

Reference RAP	Replicate	Percentage Retained (%) at Sieve Size (mm)										
		16	**13.2**	**9.5**	**4.75**	**2.36**	**1.18**	**0.6**	**0.3**	**0.15**	**0.075**	**<0.075**
1	1	1.8	4.5	21.9	35.1	17.0	6.9	5.0	4.2	1.9	0.6	1.1
	2	1.6	5.3	17.3	30.5	20.4	8.8	6.7	5.8	2.5	0.7	0.5
	3	1.8	7.5	17.1	21.8	24.6	10.9	6.5	5.7	2.6	0.9	0.4
	4	0.3	4.0	13.7	28.7	19.0	10.2	9.1	8.9	4.1	1.3	0.7
	5	1.6	5.8	13.0	31.1	19.0	9.4	7.9	7.5	3.1	1.0	0.6
2	1	2.6	2.5	12.7	30.3	24.6	10.3	7.2	5.8	2.4	0.9	0.8
	2	2.0	5.1	14.4	26.6	23.3	10.8	7.2	6.1	2.6	0.9	0.9
	3	2.2	3.9	15.1	27.1	26.4	9.4	6.6	5.1	2.3	0.9	0.9
	4	2.6	3.9	11.5	30.2	29.0	9.2	6.0	4.4	1.8	0.7	0.7
	5	1.0	5.3	17.7	24.1	22.6	10.2	7.1	6.2	3.2	1.2	1.3
3	1	8.1	12.1	20.9	28.7	13.3	5.1	4.1	4.7	1.9	0.6	0.4
	2	5.9	16.5	20.3	28.3	12.6	4.5	3.5	5.0	1.8	0.8	0.8
	3	7.3	13.6	20.9	28.2	14.2	4.8	3.9	4.3	1.8	0.6	0.5
	4	8.0	13.3	20.6	28.1	14.5	5.1	3.7	4.1	1.6	0.5	0.4
	5	7.1	14.6	20.8	27.6	14.7	4.4	3.6	4.1	1.9	0.6	0.6
4	1	14.5	17.5	17.4	17.9	18.7	4.7	3.9	3.2	1.3	0.5	0.5
	2	14.4	17.8	16.9	17.8	18.5	4.4	3.8	3.5	1.4	0.6	0.8
	3	14.6	17.8	17.0	18.1	18.6	5.1	3.7	2.9	1.1	0.4	0.5
	4	14.4	17.5	16.8	18.0	18.9	4.9	3.9	3.2	1.2	0.5	0.6
	5	14.4	17.9	16.8	18.0	18.4	4.8	3.7	3.1	1.5	0.6	0.8

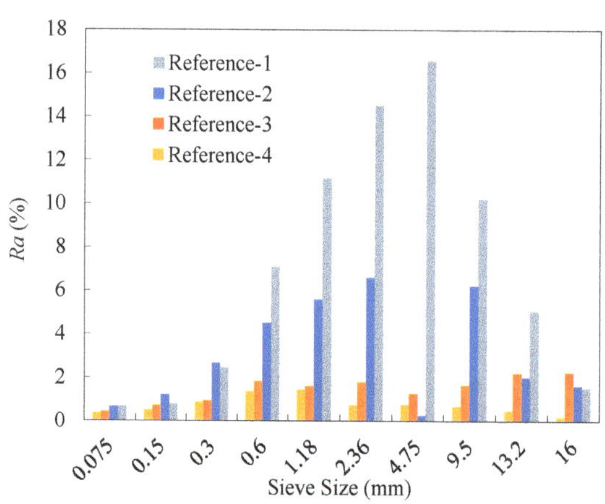

Figure 3. Ranges of the percentage retained of the four reference RAP materials.

As can be seen from Figure 4, the degree of variability in the RAP materials varies across different particle sizes for the four reference materials. Significant variability in the gradation of RAP materials is primarily observed within the size segments of 9.5~19 mm and 0~0.3 mm, whereas the variability is relatively low in the size segment of 0.3~9.5 mm. In addition, the overall variability of the RAP materials decreased as the number of fractions increased. Compared to the average *COV* value over all the sieves from the two-fraction processing, those from the four-fraction and six-fraction methods decreased by 51.5% and 73.5%, respectively.

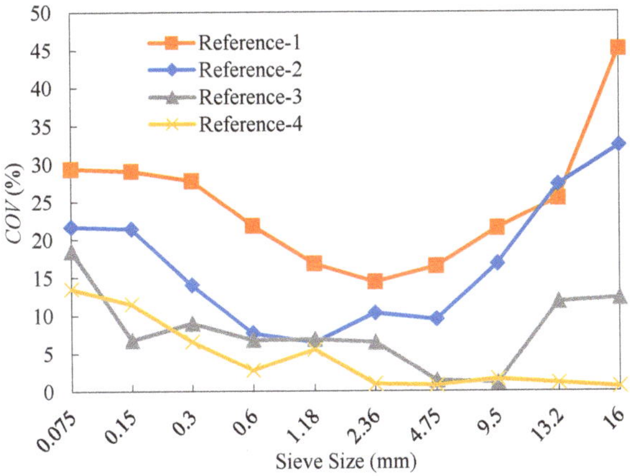

Figure 4. *COV* values of the percentage retained of the four reference RAP materials.

The old aggregates for each reference RAP material were obtained by the ignition method. The test results of percentage retained from sieve analysis are presented in Table 12. The range values of the percentage retained for the old aggregates from the four reference materials are shown in Figure 5. As can be seen, the range of the old aggregate gradation generally decreased as the number of fractions increased, demonstrating that a higher number of fractions can effectively reduce the gradation variability of the old aggregates. Figure 6 shows the distribution of *COV* values in the gradation of the old aggregates. It can be observed that the fractionation processing effectively reduced the variability of reclaimed aggregates in RAP, with the effectiveness increasing as the number of fractions rose. Compared with the average *COV* value over all the sieves from the two-fraction processing, those from the four-fraction and six-fraction methods decreased by 34.4% and 63.2%, respectively.

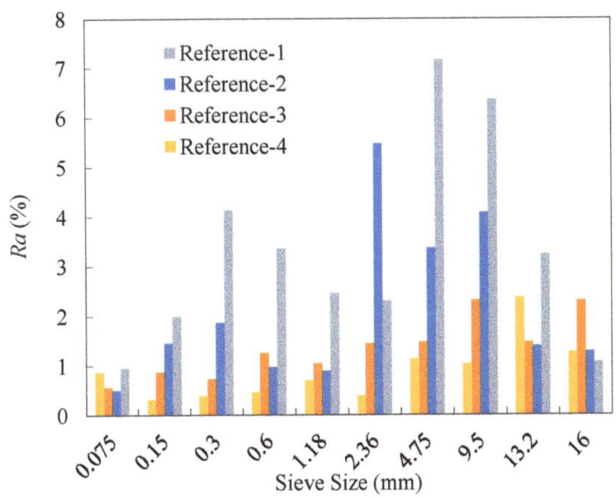

Figure 5. Ranges of the percentage retained of the old aggregates for the four reference RAP materials.

Table 12. Test results of percentage retained of old aggregates of the reference RAP materials subjected to different fractionation processing methods.

Reference Material	Replicate	Percentage Retained (%) at Sieve Size (mm)										
		16	13.2	9.5	4.75	2.36	1.18	0.6	0.3	0.15	0.075	<0.075
1	1	1.1	2.6	16.1	31.9	16.3	6.8	6.2	7.4	4.6	2.4	4.6
	2	1.1	3.3	13.2	26.8	18.6	7.7	7.3	8.9	5.3	2.8	4.8
	3	0.4	5.6	12.7	27.0	17.9	7.8	7.1	8.8	5.1	3.3	4.3
	4	0.3	2.3	10.3	24.7	17.7	9.2	9.6	11.5	6.6	2.8	5.0
	5	1.4	3.5	9.8	26.8	18.1	8.7	8.5	10.1	5.8	3.2	4.2
2	1	1.6	2.2	9.1	25.8	21.6	9.2	8.1	8.8	5.0	3.8	4.6
	2	0.8	3.4	11.2	23.3	20.8	9.7	7.6	9.2	5.3	4.2	4.4
	3	1.8	2.8	10.9	23.5	23.2	8.8	7.4	8.1	5.0	4.2	4.2
	4	2.1	2.0	8.2	25.9	25.6	8.9	7.1	7.7	4.4	3.9	4.3
	5	0.9	2.8	12.3	22.5	20.1	9.1	7.7	9.5	5.9	4.3	4.9
3	1	3.2	7.1	17.1	26.7%	14.6	6.0	5.6	8.1	4.8	3.4	3.3
	2	3.8	7.6	17.6	25.7	13.9	5.5	4.9	8.3	5.4	3.8	3.6
	3	2.7	7.1	15.9	27.1	15.0	6.1	6.2	8.0	4.8	3.7	3.4
	4	2.3	7.5	18.1	25.8	15.3	6.4	5.4	7.6	4.6	3.5	3.4
	5	4.6	8.6	15.8	26.6	15.0	5.4	5.0	7.6	4.6	4.0	2.9
4	1	6.9	11.0	16.2	18.8	17.9	6.2	5.2	6.6	4.0	3.8	3.4
	2	5.8	13.0	15.5	18.5	17.9	5.8	5.2	6.9	4.2	4.0	3.3
	3	6.6	10.8	15.6	19.6	18.1	6.4	5.5	6.6	4.1	4.6	2.1
	4	6.2	10.6	16.5	18.9	18.1	6.1	5.7	6.9	4.3	4.1	2.6
	5	5.6	12.1	15.8	19.3	17.7	5.7	5.4	6.8	4.3	4.3	3.0

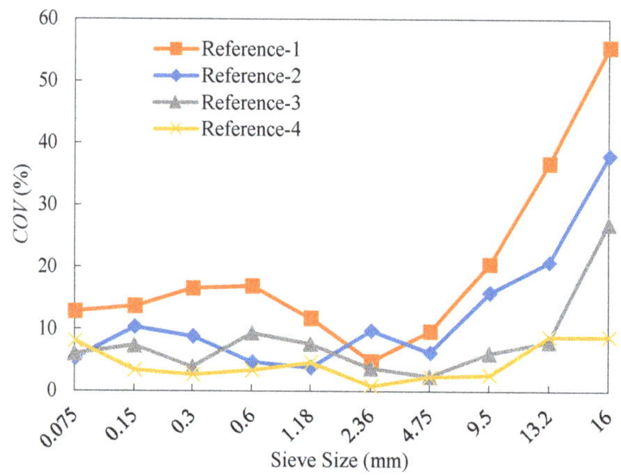

Figure 6. *COV* values of the percentage retained of the old aggregates for the four reference RAP materials.

The gradation curves of the four reference mixtures before and after the ignition test are plotted in Figure 7. It can be observed that, compared to the RAP material before ignition, the gradation of the reclaimed aggregates after being burned falls closer within the upper and lower limits specified for AC-16. Furthermore, due to long-term loading and milling operations, some coarse aggregates became fine, though the degree of fragmentation was limited. Consequently, the target gradation can still be achieved by supplementation with additional virgin coarse aggregates.

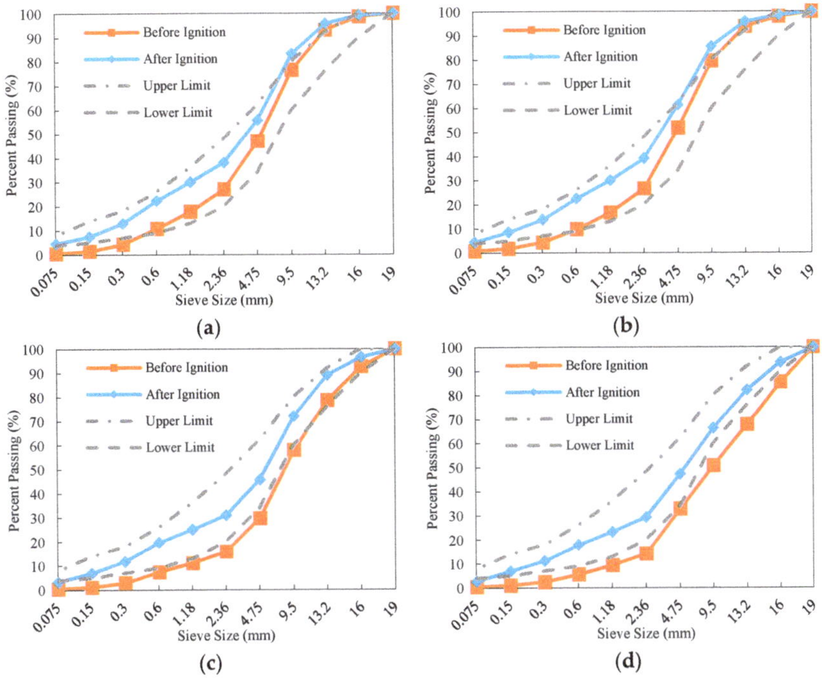

Figure 7. Grading curves of the reference RAP materials before and after ignition: (**a**) Reference-1; (**b**) Reference-2; (**c**) Reference-3; (**d**) Reference-4.

As shown in Figure 8, the tested gradations of the old aggregates in References-3 and 4 fall within the upper and lower limits specified for AC-16, while the gradation curves of References-1 and 2 for particles above 4.75 mm exceed the upper limit of AC-16. The grading curves of the old aggregates of References-1 and 2 are found very close to each other. This indicates that the widely used two-fraction method cannot effectively eliminate variability in gradation of RAP aggregates. However, as the number of fractions further increased, the gradation of the reclaimed aggregates in the RAP (see References-3 and 4) approached closer to the median of the AC-16 target gradation. This suggests that the GRA method [34] used in this study can effectively blend the fractionated RAP materials based on the particle size distribution characteristics of the reclaimed aggregates. This process significantly controls and mitigates the potential gradation variation issue of RAP materials and enables the high percentage reuse of RAP in recycled mixtures with desirable performance.

This study used RAP material from only one source, but in many cases RAP materials are collected from multiple sources. The findings from this study should be further demonstrated in multi-source scenarios. Investigation into other factors that may impact the variability of RAP properties such as particle agglomeration, asphalt aging, and binder activity is needed in future studies. Furthermore, additional research efforts are required to assess the effect of reduced RAP variability caused by refined fractionation processing on the performance of the corresponding recycled asphalt mixtures.

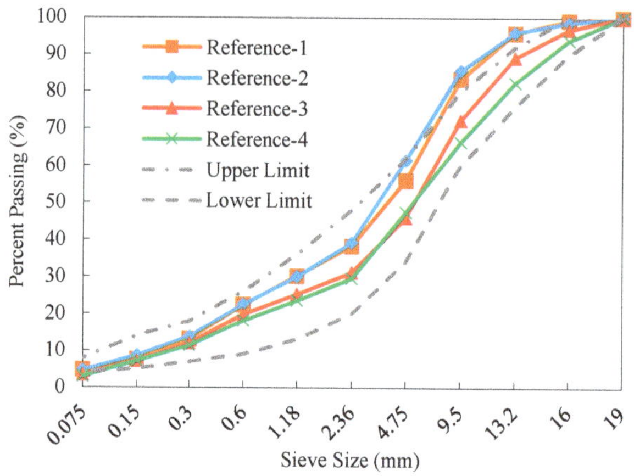

Figure 8. Tested gradations of old aggregates of the four reference RAP materials after ignition.

5. Summary and Conclusions

This study investigated approaches for evaluating and controlling variability in properties of RAP materials, including moisture content, asphalt content, and gradation. Four different fractionation methods, i.e., unfractionated, two-faction, four-fraction, and six-fraction processing approaches, were considered, and accordingly four reference RAP materials were prepared by rationally recombining the obtained RAP fractions in different proportions towards the target gradation. The GRA was adopted to identify the correlation between the gradation of old aggregates recovered from each RAP fraction and the target gradation and to determine the optimal proportions of the RAP fractions for different reference RAP materials, thus achieving a satisfactory control on heterogeneity in the gradation of the old aggregates. Various statistical indicators were adopted to assess the variability of RAP properties. Based on the results and analysis, the following conclusions can be drawn:

(1) The unfractionated method and commonly used two-fraction processing struggle to efficiently control the variability of RAP properties, particularly the gradation of old aggregates.

(2) Compared to unfractionated RAP materials, the variability of RAP materials subjected to fractionation processing showed significant improvement. Moreover, as the number of fractions increased, the degree of variability in the RAP materials remarkably decreased. Specifically, compared with the *COV* values from the commonly used two-fraction processing, those from the refined four-fraction and six-fraction processing methods declined by up to 51.5% and 73.5%, respectively.

(3) Fractionating RAP materials into four fractions in a refined manner represented the most economically viable and technically reasonable processing strategy, as it effectively controlled the variability in the properties of the RAP materials.

(4) The GRA affords a simple and effective approach for directly calculating the proportions of RAP fractions in terms of geometric similarity between grading curves without performing a nonlinear minimization algorithm on a target error function.

(5) Proportioning RAP fractions into reference materials with desirable gradations can be a potential approach for controlling the high-RAP recycled mixture with low variability and superior performance.

(6) Additional research efforts are required to investigate other factors that may cause RAP variability, like RAP source, particle agglomeration, asphalt aging, and binder activity. In addition, assessment on the effect of reduced RAP variability caused by

refined fractionation processing on the performance of the corresponding recycled asphalt mixtures is required in future studies.

Author Contributions: Conceptualization, Y.S.; Methodology, Y.Z., J.L. and Y.S.; Validation, J.L.; Investigation, J.L.; Writing—original draft, Y.Z.; Writing—review & editing, Y.S.; Funding acquisition, Y.Z. All authors have read and agreed to the published version of the manuscript.

Funding: This study was sponsored by the Fundamental Research Project of the Educational Department of Liaoning Province (LJKMZ20222050). Their support is gratefully acknowledged.

Data Availability Statement: The original contributions presented in this study are included in the article. Further inquiries can be directed to the corresponding author.

Conflicts of Interest: The authors declare no conflict of interest.

References

1. Hu, M.; Ji, S.; Sun, Y.; Zhu, K. Understanding the low-temperature fracture behavior of rejuvenated high viscosity modified asphalt utilizing a combined microstructure-component analysis. *Constr. Build. Mater.* **2025**, *471*, 140717. [CrossRef]
2. Li, M.; Yuan, J.; Jin, T.; Wang, W.; Sun, Y.; Cheng, H. Investigation of performance evolution in recycled asphalt mixtures: The impact of virgin and RAP binder blending. *Constr. Build. Mater.* **2025**, *469*, 140519. [CrossRef]
3. Xu, H.; Chen, J.; Sun, Y.; Zhu, X.; Wang, W.; Liu, J. Rheological and physico-chemical properties of warm-mix recycled asphalt mastic containing high percentage of RAP binder. *J. Clean. Prod.* **2021**, *289*, 125134. [CrossRef]
4. Katla, B.; Raju, S.; Waim, A.R.; Danam, V.A. Utilization of Higher Percentages of RAP for Improved Mixture Performance by Adopting the Process of Fractionation. *Int. J. Pavement Res. Technol.* **2021**, *15*, 349–366. [CrossRef]
5. Xu, H.; Sun, Y.; Chen, J.; Li, J.; Yu, B.; Qiu, G.; Zhang, Y.; Xu, B. Investigation into Rheological Behavior of Warm-Mix Recycled Asphalt Binders with High Percentages of RAP Binder. *Materials* **2023**, *16*, 1599. [CrossRef] [PubMed]
6. Hu, M.; Sun, D.; Hofko, B.; Sun, Y.; Mirwald, J.; Xu, L. Multiscale optimization on polymer-based rejuvenators for the efficient recycling of aged high-viscosity modified asphalt: Molecular dynamics simulation and experimental analysis. *J. Clean. Prod.* **2024**, *449*, 141736. [CrossRef]
7. Solaimanian, M.; Tahmoressi, M. Variability Analysis of Hot-Mix Asphalt Concrete Containing High Percentage of Reclaimed Asphalt Pavement. *Transp. Res. Rec. J. Transp. Res. Board* **1996**, *1543*, 89–96. [CrossRef]
8. Li, M.; Han, Z.; Cheng, H.; Yang, R.; Yuan, J.; Jin, T. Low-temperature performance improvement strategies for high RAP content recycled asphalt mixtures: Focus on RAP gradation variability and mixing process. *Fuel* **2025**, *387*, 134362. [CrossRef]
9. Xu, G.; Shen, Z.; Hao, S.; Du, M.; Hu, J. Mitigating RAP variability through detailed particle size classification: Applications in high RAP content asphalt mixtures. *Case Stud. Constr. Mater.* **2025**, *23*, e04993. [CrossRef]
10. Montañez, J.; Caro, S.; Carrizosa, D.; Calvo, A.; Sánchez, X. Variability of the mechanical properties of Reclaimed Asphalt Pavement (RAP) obtained from different sources. *Constr. Build. Mater.* **2020**, *230*, 116968. [CrossRef]
11. Austerman, A.J.; Mogawer, W.S.; Stuart, K.D. Variability of Reclaimed Asphalt Pavement (RAP) Properties within a State and Its Effects on RAP Specifications. *Transp. Res. Rec. J. Transp. Res. Board* **2020**, *2674*, 73–84. [CrossRef]
12. Aurangzeb, Q.; Al-Qadi, I.L.; Abuawad, I.M.; Pine, W.J.; Trepanier, J.S. Achieving Desired Volumetrics and Performance for Mixtures with High Percentage of Reclaimed Asphalt Pavement. *Transp. Res. Rec. J. Transp. Res. Board* **2012**, *2294*, 34–42. [CrossRef]
13. Wang, J.; Xu, J.; Liu, L. Variability Investigation of Reclaimed Asphalt Pavement Materials. *J. Mater. Civ. Eng.* **2024**, *36*, 04023558. [CrossRef]
14. Hong, F.; Guo, R.; Zhou, F. Impact of recycled asphalt pavement material variability on pavement performance. *Road Mater. Pavement Des.* **2014**, *15*, 841–855. [CrossRef]
15. Pan, Y.; Li, J.; Yang, T.; Liu, G.; Zhou, J.; Guo, P.; Zhao, Y. Optimization of gradation design of recycled asphalt mixtures based on fractal and Mohr-Coulomb theories. *Constr. Build. Mater.* **2020**, *248*, 118649. [CrossRef]
16. Rema, A.; Swamy, A.K. Response Surface–Based Approach to Quantify Variability in Recycled-Asphalt Concrete Mixtures. *J. Mater. Civ. Eng.* **2022**, *34*, 04022315. [CrossRef]
17. Zaumanis, M.; Oga, J.; Haritonovs, V. How to reduce reclaimed asphalt variability: A full-scale study. *Constr. Build. Mater.* **2018**, *188*, 546–554. [CrossRef]
18. Zaumanis, M.; Boesiger, L.; Kunz, B.; Mazzoni, H.; Bruhin, P.; Mazor, S.; Poulikakos, L. Three indexes to characterise crushing and screening of reclaimed asphalt pavement. *Int. J. Pavement Eng.* **2021**, *23*, 4977–4990. [CrossRef]
19. R., A.; Radhakrishnan, V.; Bharath, G. Effect of RAP Dissociation and Agglomeration on Recycled Asphalt Binder Availability. *J. Mater. Civ. Eng.* **2024**, *36*, 04024264. [CrossRef]

20. He, J.; Ma, Y.; Zheng, K.; Cheng, Z.; Xie, S.; Xiao, R.; Huang, B. Quantifying the agglomeration effect of reclaimed asphalt pavement on performance of recycled hot mix asphalt. *J. Clean. Prod.* **2024**, *442*, 141044. [CrossRef]
21. Ferreira, W.L.G.; Castelo Branco, V.T.F.; Vasconcelos, K.; Bhasin, A.; Sreeram, A. The impact of aging heterogeneities within RAP binder on recycled asphalt mixture design. *Constr. Build. Mater.* **2021**, *300*, 124260. [CrossRef]
22. Sreeram, A.; Leng, Z. Variability of rap binder mobilisation in hot mix asphalt mixtures. *Constr. Build. Mater.* **2019**, *201*, 502–509. [CrossRef]
23. Zhu, J.; Ma, T.; Fang, Z. Characterization of agglomeration of reclaimed asphalt pavement for cold recycling. *Constr. Build. Mater.* **2020**, *240*, 117912. [CrossRef]
24. Gao, J.; Yang, J.; Yu, D.; Jiang, Y.; Ruan, K.; Tao, W.; Sun, C.; Luo, L. Reducing the variability of multi-source reclaimed asphalt pavement materials: A practice in China. *Constr. Build. Mater.* **2021**, *278*, 122389. [CrossRef]
25. De Lira, R.R.; Cortes, D.D.; Pasten, C. Reclaimed asphalt binder aging and its implications in the management of RAP stockpiles. *Constr. Build. Mater.* **2015**, *101*, 611–616. [CrossRef]
26. Yu, X.; Tang, W.; Li, N.; Jiang, M.; Huang, J.; Wang, D. Refined decomposition: A new separation method for RAP materials and its effect on aggregate properties. *Constr. Build. Mater.* **2022**, *358*, 129452. [CrossRef]
27. Wang, P.; Chen, J.; Wang, J.; Li, J.; Ning, H.; Liang, C.; Ge, X.; Wang, X. Evaluation of the Refined Decomposition Effect of Reclaimed Asphalt Pavement Materials. *Buildings* **2023**, *13*, 2240. [CrossRef]
28. Shannon, C.; Mokhtari, A.; Lee, H.D.; Tang, S.; Williams, C.; Schram, S. Effects of High Reclaimed Asphalt-Pavement Content on the Binder Grade, Fatigue Performance, Fractionation Process, and Mix Design. *J. Mater. Civ. Eng.* **2017**, *29*, 04016218. [CrossRef]
29. Nanda, H.; Siddagangaiah, A.K. Effect of RAP fractionation and dosage on design and mechanical behaviour of cold asphalt mixes. *Constr. Build. Mater.* **2024**, *422*, 135773. [CrossRef]
30. Shen, J.; Amirkhanian, S.; Miller, J.A. Effects of Rejuvenating Agents on Superpave Mixtures Containing Reclaimed Asphalt Pavement. *J. Mater. Civ. Eng.* **2007**, *19*, 376–384. [CrossRef]
31. *JTG 3432-2024*; Test Methods of Aggregates for Highway Engineering. China Communications Press: Beijing, China, 2024.
32. *ASTM D5404/D5404M*; Standard Practice for Recovery of Asphalt Binder from Solution Using the Rotary Evaporator. ASTM International: West Conshohocken, PA, USA, 2024.
33. *JTG/T 5521-2019*; Technical Specifications for Highway Asphalt Pavement Recycling. China Communications Press: Beijing, China, 2019.
34. Liu, S. *Grey Systems Analysis: Methods, Models and Applications*; Springer: Singapore, 2025.
35. Guo, K. The Study on Pavement Performance of the Different Mixing Rate Regeneration Mixture Based on Reference Material Synthesis. Chongqing Jiaotong University, Chongqing, China, 2013.
36. *ASTM D6307*; Standard Test Method for Asphalt Content of Asphalt Mixture by Ignition Method. ASTM International: West Conshohocken, PA, USA, 2019.

materials

MDPI

Article

New Type of Asphalt Concrete with Bitumen Emulsion in Terms of Type and Quantity of Binder

Maciej Krasowski [1,*], Przemysław Buczyński [1], Grzegorz Mazurek [1] and Matúš Kozel [2]

1 Department of Transportation Engineering, Faculty of Civil Engineering and Architecture, Kielce University of Technology, Al. Tysiąclecia Państwa Polskiego 7, 25-314 Kielce, Poland; p.buczynski@tu.kielce.pl (P.B.); gmazurek@tu.kielce.pl (G.M.)

2 Department of Technology and Construction Management, Faculty of Civil Engineering, University of Žilina, 8215/1, 01026 Zilina, Slovakia; matus.kozel@uniza.sk

* Correspondence: krasowski4.mk@gmail.com; Tel.: +48-785-076-740

Abstract

This paper presents the effect of modifiers on the properties of a mixture of asphalt concrete with bitumen emulsion (ACBE). The mineral-asphalt mixture is the only one that can be produced using the cold-mix technology (CMA). The theoretical part of the article details the characteristics of the methods for producing mineral-asphalt mixtures in terms of their production temperature. Thus, hot (HMA), half-warm (H-WMA), warm (WMA) and cold (CMA) mixtures are discussed. The research section presents the design of the asphalt concrete composition with bitumen emulsion, the research methods, the experiment design and the research results. The design of the mixture of asphalt concrete with bitumen emulsion was carried out in accordance with the guidelines set out in EN 13108-31. In the experiment, Portland cement (C), bitumen emulsion (A), synthetic latex (styrene-butadiene rubber SBR) (B) and redispersible polymer powder EVA (polyethylene-co-vinyl acetate) (P) were used as modifiers. Twenty-four mixtures were designed as part of the experiment, according to the 3^4 experiment design. The following physical and mechanical properties were assessed in the design of the research: air void content V_m, water ab-sorption n_w, indirect tensile strength ITS and IT-CY stiffness modulus. When analysing the research results, the authors observed a noticeable impact of the content of asphalt (A) and synthetic latex (B) on the air void content V_m. A significant effect was also observed for the interaction of Portland cement (C) and redispersible polymer powder (P) on the indirect tensile strength ITS. The next step was the optimisation of the ACBE mixture composition, which effect made it possible to identify the optimum amounts of modifiers in the mixture of asphalt concrete with bitumen emulsion (ACBE), which constituted recommendations for the requirements for mixtures of asphalt concrete with bitumen emulsion.

Keywords: cold mix asphalt; polymer powder; design of experiment; bitumen emulsion

1. Introduction

Pavements made from mineral-asphalt mixtures have been used successfully throughout the world for many years. Over the years, production processes have been refined through various modifications [1]. A major step in the development of asphalt production methods was the commencement of modifications using natural and synthetic polymers. In the early 1980s, a styrene-butadiene-styrene (SBS-type) polymer was produced for the first time [2]. In Europe and the USA, the use of polymer-modified bitumens began in the second half of the 1980s [2]. Polymer-modified bitumens appeared in Poland in the 1990s [3].

Over the years, mixture production methods and technologies have evolved as well. The oldest and most common technology is the production of hot-mix asphalt (HMA) [4]. This technology involves heating the mineral aggregate and asphalt separately. The aim is to reduce the viscosity of the asphalt (liquefaction), and heating the aggregate helps to coat the aggregate grains with the binder [4]. Adhesive agents are added to facilitate and improve adhesion. The production temperature of the mixture is between 150 °C and 190 °C [4]. Due to high production temperatures, it is believed that HMA mixtures are not beneficial for the environment [5]. This is associated with significant atmospheric emissions of gases. HMA production is also very energy-intensive [6,7].

A more environmentally friendly mixture with a lower production temperature than in the case of HMA is the "warm" mix asphalt. This technology has been developed since the early 1990s, mainly in the USA [8]. Eventually, it emerged in Europe. This method involves reducing the viscosity of the asphalt without heating it to high temperatures, like HMA [9]. One way of performing this is to use suitable additives such as waxes or fatty acid amides. These chemicals are usually dosed at a rate of 0.1–3% of asphalt weight in the mixture. Another way to lower the viscosity of asphalt is to use the foamed bitumen technology [10]. HMA technology involves adding water and pressurised air to the hot asphalt. The hot asphalt (at approximately 160–180 °C) is poured into a special system where water and compressed air can be added. Then, a small amount of water is injected into the hot asphalt line. As a result, the water rapidly evaporates, forming tiny bubbles surrounded by a thin bitumen film. The resulting foamed bitumen is mixed with mineral aggregate [11]. The production temperature of the WMA mix is approximately 100–150 °C [12,13].

Another type of mineral-asphalt mixture is the "half-warm" mix asphalt (H-WMA). This technology was invented in 1999 and is partly classified as a WMA mix. That is due to the bitumen foaming technology involved [14]. Using a suitable method of bitumen foaming, it is possible to pour mixtures at a temperature of 60–100 °C. An example of the technology used is Evotherm ET [15]. The method involves heating the mineral aggregate to 120–130 °C and then adding a bitumen emulsion with the polymer modifier SBR (styrene–butadiene–rubber). The added ingredients initiate the bitumen foaming process [16]. The final mixture has a temperature of 90–100 °C [17,18].

The last of the technologies is the cold mix asphalt (CMA). This method involves producing mixtures at temperatures in the 0–30 °C range. This is possible by using bitumen emulsion as a binder [19]. The bitumen emulsion is added to the wet mineral mix without heating. Upon contact between the aggregate and the emulsion, the emulsion breaks down. As a result of compaction, water is removed from the mixture, and the dispersed bitumen forms bonds between the grains of the mineral material [20,21]. This method is successfully used worldwide [22–24]. The following figure (Figure 1) shows a breakdown of mineral-asphalt mixtures according to the temperature of production.

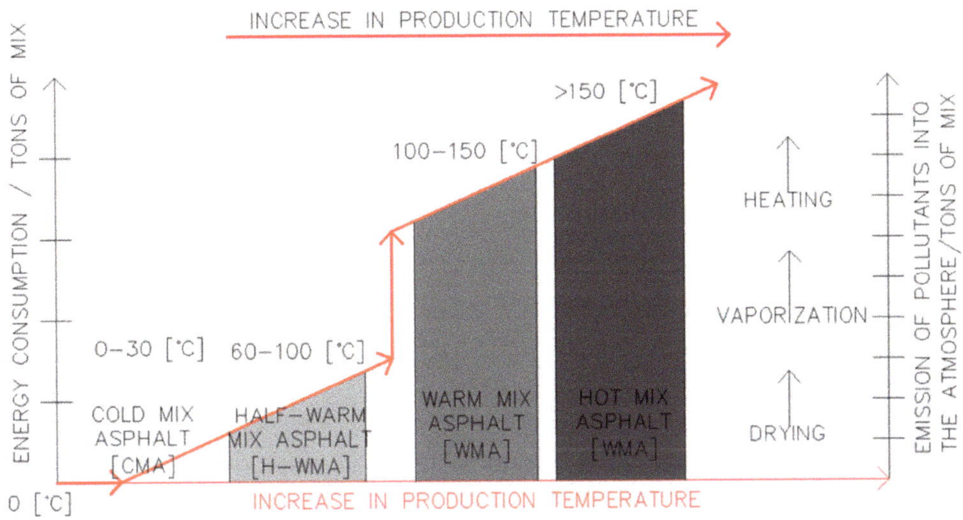

Figure 1. A breakdown of mineral-asphalt mixtures according to the temperature of production [25].

The CMA method has many advantages. An example is the low production temperature, which translates into lower energy requirements, resulting in lower emissions of harmful gases. Additionally, bitumen emulsions may contain additives (e.g., reinforcing fibres) [26] that remain after water evaporation, yielding composite asphalt with improved mechanical properties. What is more, the low production temperature translates into longer times available for the transport and pouring of the mixture. It is also possible to commission the completed layer more quickly than for the conventional HMA mixture. Despite its many advantages, the mix also has disadvantages. The main disadvantage cited by researchers is the high air void content, Vm, and low frost and water damage resistance [27,28].

Consequently, it seems necessary to look for methods to improve the CMA technology. Current global trends aimed at reducing environmental pollution [29] are forcing the use of environmentally friendly materials that generate low atmospheric emissions. The CMA mix seems to be the perfect starting point for further development. The introduction of modifiers may prove to be a way of improving some of the properties for which the conventional CMA technology is not chosen by contracting authorities and contractors. It is necessary to look for the optimum amount of binders to produce a new type of asphalt concrete with bitumen emulsion.

The subject of the research is the mineral-asphalt mixture produced using the CMA technology. The authors have selected the mixture of asphalt concrete with bitumen emulsion (ACBE) for the analysis. According to the definition [30], an ACBE mix is a mix in which a continuous mineral skeleton forms an interlocking structure in which all or part of the binder is added in the form of bitumen emulsion.

2. Materials and Methods

2.1. Materials

The design of the mixture of asphalt concrete with bitumen emulsion began with the correct design of the mineral skeleton. Crushed mineral aggregate from a quartzite aggregate mine was used in the mix. Thus, the designed mineral mix [MM] had a grain size consistent with the requirements of [30]. The asphalt concrete mixture is continuously graded. The following figure (Figure 2) shows the grain size curve.

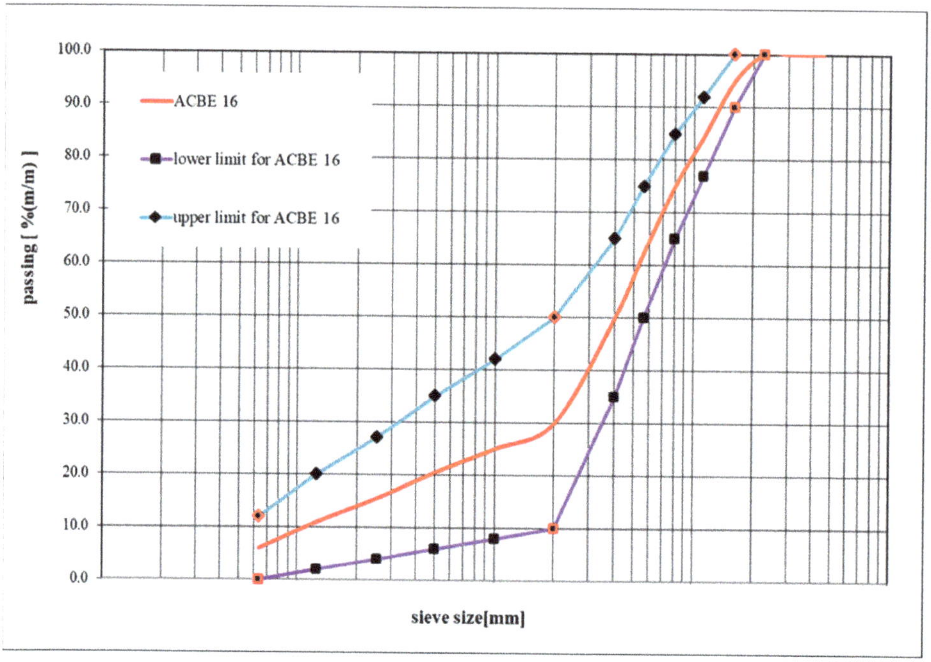

Figure 2. Gradation curve of the ACBE mineral mix.

The next stage of the ACBE mix design process involved designing the amount of binders. In line with the concept, the authors provided for the use of four binders in the mix, i.e., bitumen emulsion (A), multicomponent Portland cement (C), synthetic latex (B), and redispersible polymer powder (RPP). As a result of the assumptions, the final composition of the MCE mix was as shown in Table 1.

Table 1. Percentage of mineral mix ingredients.

Ingredient	Origin	Density [g/cm^3]	MM [%, m/m]
0/2 fine aggregate	Quartzite	2.68	35.00–38.80
2/5 quartzite chippings	Quartzite	2.63	21.00–23.30
5/8 quartzite chippings	Quartzite	2.66	10.50–11.60
8/11 quartzite chippings	Quartzite	2.66	10.50–11.60
11/16 quartzite chippings	Quartzite	2.65	10.50–11.60
Bitumen emulsion	-	1.00	3.33–10.00
Multicomponent Portland cement	-	2.99	0.00–2.25
Synthetic latex	-	1.00	0.00–6.00
Redispersible polymer powder	-	0.50	0.00–3.00

The composition of the mixture of asphalt concrete with bitumen emulsion (ACBE) was designed in accordance with the guidelines of EN 13108-31 [30] in terms of the limit curves. A grain size of up to 16 mm was adopted. The mineral mix forming the skeleton of ACBE mixtures consisted of quartzite chippings with a grain size of 0/2 at 40% (m/m), 2/5 at 24% (m/m), 5/8 at 12% (m/m), 8/11 at 12% (m/m) and 11/16 at 12% (m/m). Table 2 below shows the properties of the aggregates used in the composition of the mineral mix.

Table 2. Properties of the aggregates used in the project.

Property	Test	Unit of Measure	Symbol				
Size	EN 933-1 [31]	-	0/2	2/5.6	5.6/8	8/11	11/16
Grain size	EN 933-1 [31]	-	G_F85	Gc85/20	Gc85/20	Gc90/15	Gc85/35
Density	EN 1097-6 [32]	Mg/m^3	2.68	2.63	2.66	2.66	2.65
Flakiness index	EN 933-3 [33]	%	-	Fl_{20}	Fl_{15}	Fl_{15}	Fl_{10}
Percentage of crushed or broken grains	EN 933-5 [34]	%	-	$C_{100/0}$	$C_{100/0}$	$C_{100/0}$	$C_{100/0}$
Frost resistance	EN 1367-1 [35]	%	-	F_1	F_1	F_1	F_1
Crushing resistance of coarse aggregate	EN 1097-2 [36]	%	-	LA_{25}	LA_{25}	LA_{25}	$LA1_5$
Resistance to abrasion	EN 1097-1 [37]	%	-	$M_{DE}10$	$M_{DE}10$	$M_{DE}10$	$M_{DE}10$

The production of a mixture of asphalt concrete with bitumen emulsion requires an appropriate approach in terms of composition design. An attempt was made to identify the optimum proportion of binders in order to ensure adequate performance of the finished mix. The authors used the following in the project:

- Bitumen emulsion (A);
- Portland cement (C);
- Synthetic latex (B);
- Redispersible polymer powder (P).

ACBE mixtures contain asphalt, which is an ingredient of bitumen emulsion. Bitumen emulsion is added to ACBE mixtures, and when mixed with the mixture components, it breaks down into bitumen and water with additives. Asphalt from the bitumen emulsion coats the aggregate grains, acting as a binder in the mix.

A 70/100 viscosity road bitumen was used to produce the binder. The asphalt emulsion meets the requirements of the national annex to the standard for cationic asphalt emulsions—PN-EN 13808 [38] and is designated C60B10R.

According to the definition of the standard [38], this designates a cationic asphalt emulsion (C) with an emulsion asphalt content of 60% (60), produced from road bitumen (B), with a degradation class of 10 (10), intended for recycling (R).

Table 3 summarises the parameters characterising the asphalt emulsion used.

Table 3. Properties of the C60B10 R bitumen emulsion [39].

Property	Test	Unit of Measure	Value
Binder content	PN-EN 13808 [38]	% (m/m)	60.0
Stability of mixing with cement		g	0.3
Percent retained on 0.5 mm sieve		% (m/m)	0.06
Ø2 mm efflux time at 40 °C		s	27
Asphalt-aggregate bond strength		%	75
Reclaimed asphalt penetration grade		0.1 mm	53
Softening point of reclaimed asphalt		°C	55.2

The amount of emulsifier in the emulsion composition was 1.2%. The size of dispersed asphalt particles ranged from 1 to 20 microns, with the largest share being particles of 4 microns.

The mixtures were prepared using fly ash Portland cement of class II, with a strength of 42.5 MPa and with a high early strength "R", as determined in accordance with EN 197-1 [40]. The main properties of the hydraulic binder used are shown below in Table 4.

Table 4. Properties of CEM II 42.5R Portland cement.

Property	Test	Unit of Measure	Value
Initial setting time	EN 196-3 [41]	min	246
Compressive strength at 2 days	EN 196-1 [42]	MPa	26.2
Compressive strength at 28 days	EN 196-1 [42]	MPa	57.5
Soundness	EN 196-3 [41]	mm	0.8
Specific surface area	EN 196-6 [43]	cm2/g	4354

Another ingredient used in the project was the modifier for cationic bitumen emulsions—synthetic latex in liquid form. Synthetic latex is a mixture of styrene-butadiene copolymer (styrene-butadiene dispersion) in the amount of 64 (%; m/m) and water together with water-soluble components, e.g., HCl in the amount of 34 (%; m/m). It was dosed into the bitumen emulsion as a binder modifier. Mixtures of asphalt concrete with bitumen emulsion have been made with the addition of a polymer belonging to the plastomer group [44]. The polymer used is a thermoplastic EVA (polyethylene-co-vinyl acetate) copolymer. It comes in the form of a white powder formed by the evaporation of water from the polymer dispersion. It is obtained in the spray-drying process [39]. When mixed with water, this powder forms a dispersion. An important advantage of this product is that it can be mixed with other dry components, such as cement. The mixture thus created produces a cement-polymer binder.

The redispersible polymer powder introduced into the concrete mix forms a dispersion with the liquid phase of the cement grout acting as the dispersed phase. Polymer base: vinyl acetate—ethylene copolymer (EVA/VAE). Protective colloid: polyvinyl alcohol (PVA). Additives: mineral anti-caking agents (calcium carbonate ($CaCO_3$)). Due to the start of the hydration process and the loss of water in the mixture, the polymer particles move closer together, forming a closely packed structure and, in the next phase, a continuous layer. A key element in the formation of a continuous polymer phase is coalescence, a phenomenon during which the particles of the dispersed phase are bound together. The phenomenon leads to a reduction in the interfacial surface, which is a very favourable phenomenon [45]. The chemical composition of the redispersible polymer powder is given in Table 5.

Table 5. Chemical composition of the EVA polymer [46].

Component	Participation [%]
C	67.67
O	29.13
Mg	0.52
Si	1.65
Ca	0.75
Al	0.29

According to an analysis of the studies [44,46–48], the use of the above-mentioned polymer modifier can improve the mechanical properties of the mixture. This is caused by the occurrence of cross-links between the particles, which produce a continuous polymer phase [44].

The materials described were used to prepare the ACBE mixture with additives. In order to obtain a thoroughly homogeneous mixture, the ingredients were mixed in a WIRTGEN WLM 30 laboratory mixer. Each time, 30 kg of the ingredients was weighed, which was an amount that allowed for thorough mixing.

The specimens made in the laboratory were compacted according to the requirements of the test method [30], i.e., Marshall specimens were compacted statically, using a hydraulic press capable of maintaining constant pressure. The specimens were compacted by applying a load corresponding to 11.9 MPa, maintained continuously for 300 s.

The amount of the material was selected in such a way that the specimen after compaction had a height of 63.5 mm (+/− 2 mm) in accordance with the standard. The specimens prepared this way on the first day after production were stored at room temperature +20 ± 5 °C, on perforated base stands to guarantee uniform drying.

2.2. Scope Research and Experiment Design

The scope of research was to verify the effect of binders on the parameters of the ACBE mix. The variables were controlled in the experiment to identify their effect on the properties of the ACBE mixture. The variables included bitumen emulsion (A), Portland cement (C), synthetic latex (styrene–butadiene rubber SBR) (B), and redispersible polymer powder EVA (polyethylene–co-vinyl acetate) (P). ACBE mixtures exhibit characteristics similar to those of cold-recycled deep-cycle mixtures. The authors have experience using redispersible polymer powder and cement [30–33] in this type of mixture. Additionally, the use of synthetic latex was intended to simulate the use of an emulsion modified with SBR synthetic latex. The analysis of the impact of binders was based on a 34 design modified using the A-optimal optimisation algorithm. In accordance with the experiment design, each of the controlled variables had a different content. Asphalt was added to the ACxBE mix at a rate of 2.00% to 6.00% with an interval of 1.00%. Multicomponent Portland cement was dosed at 0.00% to 2.25% with a 0.75% interval. Synthetic latex was used at a rate of 0.00% to 6.00% with an interval of 2.00%. Redispersible polymer powder was added at a rate of 0.00% to 3.00% with an interval of 1.00%. To assess the properties of the ACBE mix, the authors used well-known test methods commonly used in the testing of mineral-asphalt mixtures. Table 6 shows the scope of laboratory tests.

Table 6. Scope of research.

Property	Test Standard
Air void content (V_m)	PN-EN 12697-8 [49]
Water absorption by weight (n_w)	PN-S 04001-12 [50]
Indirect tensile strength (ITS)	PN-EN 12697-23 [51]
Stiffness modulus S_m at 13 °C	PN-EN 12697-26 [52]

The initial experiment design used to assess the impact of the modifiers was 34 designs. The design provides for independent variables of a qualitative as well as quantitative type at three levels. Implementing the full design (Response Surface Methodology, denoted as RSM) would require enormous amounts of materials and time. For this task, adopting a second-order response-surface regression model in the form below was needed—Formula (1):

$$Y = \beta_0 + \sum_{i=1}^{4} \beta_i X_i + \sum_{i=1}^{4} \beta_{ij} X_i X_j + \sum_{i<j} \beta_{ij} X_i X_j \quad (1)$$

where Y is the response variable (Vm, nw, ITS_{dry}, ITSR, or Sm), and X_1, X_2, X_3, X_4 are the coded levels of the four factors: (1) A—asphalt content [%], (2) C—Portland cement content [%], (3) B—synthetic latex content [%], (4) P—redispersible polymer powder content [%].

In this case, the number of combinations would be 81. The initial design was modified using an A-optimal optimisation algorithm. The goal was to construct a new experiment design that would provide for the following:

- excluding cases that were not feasible (extreme cases in the experiment design),
- considering the economic factors (the budget of the research programme),
- minimising the error of the estimation of the future response surface.

The A-optimal design provides many options to select from a list of acceptable points, i.e., design layouts that, for the selected model, would provide the maximum amount of information about the researched area. It is therefore necessary to create a list of proposed points, indicate the model to be fitted to the measurement results and specify the required number of design layouts. The algorithm used allowed for the creation of a new plan consisting of the required number of layouts, with the columns of the design matrix still orthogonal to each other to the maximum possible extent. The rationale for A-optimality is discussed, for example, in Box and Draper [53]. Due to the required calculations, the update of the matrix trace (for the A-optimality criterion) is a slow and computationally demanding process. The A-optimality criterion, characterised by the minimisation of the trace of the inverse of the X′X matrix, was selected for the optimisation process [54]. The maximum value of the following criterion (G-optimality) was used as an optimisation criterion—Formula (2).

$$G_{optimal} = 100 \times \frac{\sqrt{\frac{p}{N}}}{\Sigma M} \quad (2)$$

where p—number of effects, N—number of required layouts, ΣM—maximum standard deviation of the predicted value of the dependent variable, including all proposed points.

A G-optimal design is defined as one that minimises the largest standard deviation of the determinable response surface. This task was performed using Statistica (version 13.3) software (with the DoE module). After an iterative process taking into account the initial conditions and the objective function of the optimisation process, the authors identified 23 significant cases (reduced from 81 for a full factorial design) to predict the model with the lowest aberration for the assumption of maximising the response satisfying the orthogonality postulate. The optimisation process was closely related to the previously adopted model of the second-degree polynomial, which takes into account the presence of independent factors and the interactions between them. In this case, it is possible to determine a non-linear response surface function in the form of a polynomial of degree two.

The authors have also prepared a reference mix, designed as asphalt concrete with bitumen emulsion, for a binder course with a grain size of up to 16 mm, the same as that specified in the project. The mix contains only 4.75% of asphalt in its composition. The mineral structure of the reference mix was designed in concert with the other mixes used in the project. The asphalt content in the mix was determined using the Duriez Method [55]. This method is based on determining the specific surface area of the aggregate used in the designed mix.

In accordance with the experiment design, each of the controlled variables had a different content. The experimental domain is shown in Table 7 below.

Table 7. Experiment domain.

	MIN [%]	MAX [%]	STEP [%]
Asphalt (A)	2.00	6.00	2.00
Portland Cement (C)	0.00	2.25	0.75
Synthetic Latex (B)	0.00	6.00	2.00
Pedispersible polymer powder (P)	0.00	3.00	1.00

The number of modifiers used has been selected on the basis of literature analysis, own experience, and requirements applicable in Poland [30]. Table 8 delineates the experimental design as a second-order Response Surface Methodology (RSM) employing an A-optimal design alongside regression-based coefficient estimation.

Table 8. Experiment design for the evaluation of the impact of redispersible polymer powder on the properties of the MCE mix using RSM employing an A-optimal design alongside regression-based coefficient estimation.

Type of Mix ACBE	A (Asphalt) [%]	C (Portland Cement) [%]	B (Synthetic Latex) [%]	P (Redispersible Polymer Powder) [%]
A3C0B0P0	3.00	0.00	0.00	0.00
A3C0B0P2	3.00	0.00	0.00	2.00
A3C0B4P0	3.00	0.00	4.00	0.00
A3C0B4P2	3.00	0.00	4.00	2.00
A3C1.5B0P0	3.00	1.50	0.00	0.00
A3C1.5B0P2	3.00	1.50	0.00	2.00
A3C1.5B4P0	3.00	1.50	4.00	0.00
A3C1.5B4P2	3.00	1.50	4.00	2.00
A5C0B0P0	5.00	0.00	0.00	0.00
A5C0B0P2	5.00	0.00	0.00	2.00
A5C0B4P0	5.00	0.00	4.00	0.00
A5C0B4P2	5.00	0.00	4.00	2.00
A5C1.5B0P0	5.00	1.50	0.00	0.00
A5C1.5B0P2	5.00	1.50	0.00	2.00
A5C1.5B4P0	5.00	1.50	4.00	0.00
A5C1.5B4P2	5.00	1.50	4.00	2.00
A2C0.75B2P1	2.00	0.75	2.00	1.00
A6C0.75B2P1	6.00	0.75	2.00	1.00
A4C2.25B2P1	4.00	2.25	2.00	1.00
A4C0.75B6P1	4.00	0.75	6.00	1.00
A4C0.75B2P3	4.00	0.75	2.00	3.00
A4C0.75B2P1	4.00	0.75	2.00	1.00
A4C0.75B2P1	4.00	0.75	2.00	1.00
A4.75-REF	4.75	0.00	0.00	0.00

To facilitate the analysis and description of the research work, all mixtures were designated using a code Formula (3). The capital letter of the alphabet (A, C, B, P) indicates the ingredient, while the number indicates the percentage of each ingredient in the mixture. For example, the mixture code reads:

$$\text{A4C0.75B6P1} \tag{3}$$

where A4—asphalt content in the ACBE mix, at 4.00% (m/m); C0.75—multicomponent Portland cement content in the ACBE mix, at 0.75% (m/m); B6—synthetic latex content in the ACBE mix, at 6.00% (m/m); P1—redispersible polymer powder content in the ACBE mix, at 1% (m/m).

2.3. *Research Methods*

2.3.1. Air Void Content

The first parameter to be analysed is the air void content (Vm) defined by EN 12697-8 [49] as the volume of air voids in the specimen expressed as a percentage of the total volume of the specimen. This dependence is described according to Formula (4).

$$V_m = \frac{\rho_m - \rho_b}{\rho_m} * 100\% \quad (4)$$

where V_m—air void content [0.1%], ϱ_m—density of the mineral-asphalt mix [Mg/m^3], ϱ_b—bulk density of the mineral-asphalt mix [Mg/m^3] [56]

2.3.2. Water Absorption by Weight

Another assessed parameter is the water absorption of specimens of asphalt concrete with bitumen emulsion. Water absorption (nw) is the mass and volume of water absorbed by a specimen immersed in water for 24 h at +25 ± 5 °C. After that time, the specimen is removed from the water and dried to constant weight. Water absorption by weight (nw) is calculated as a percentage (m/m) with an accuracy of 0.1% using Formula (5) [57]:

$$n_w = \frac{m_1 - m}{m} * 100\% \quad (5)$$

where n_w—water absorption by weight [0.1%], m_1—weight of the specimen saturated with water [g], m—weight of the dry specimen [g],

2.3.3. Indirect Tensile Strength ITS_{DRY}

As part of the research work, the authors tested the indirect tensile strength (ITSDRY) of the ACBE mixture. The analysis was carried out on Marshall specimens with a diameter of 101.6 ± 0.3 mm and a height of 62.5 ± 2.5 mm. The test was performed at 25 °C ± 2 °C. The test is carried out by placing the specimens between two plates and subjecting them to compression with a constant displacement rate of 50 ± 2 mm/min. The indirect tensile strength ITSdry is calculated according to Formula (6).

$$ITS_{DRY} = \frac{2 \cdot P}{\pi \cdot h \cdot D} \quad (6)$$

where P—maximum failure load of the specimen [N], h—height of the specimen [mm], D—diameter of the specimen [mm].

2.3.4. Water and Frost Resistance ITSR

The authors assessed the resistance to climatic conditions. The test concerned the ITSR index, which allows for an assessment of the impact of water and frost. This parameter is a standard factor describing the strength of a mineral-asphalt mixture [58]. Consequently, it was implemented for the mixture of asphalt concrete with bitumen emulsion. Specimens for the study were conditioned over a period of 28 days after compaction. After this time, the specimens provided for testing were subjected to conditioning in water and a freeze–thaw cycle. The test was then carried out at 25 °C ± 1 °C, and the result of the ITSR parameter was calculated using Formula (7).

$$ITSR = \frac{ITSR_{WET}}{ITS_{DRY}} \quad (7)$$

where ITSR—frost and water resistance [%], $ITSR_{WET}$—indirect tensile strength of the conditioned specimens at 25 °C [kPa], ITS_{DRY}—indirect tensile strength at 25 °C of unconditioned specimens [kPa].

2.3.5. Stiffness Modulus

The stiffness modulus is an extremely important parameter for mixtures used in road pavement courses. To determine this parameter in the ACBE mixture, a test was performed in the IT-CY indirect tensile scheme. The test was carried out in accordance with PN-EN 12697-26 [52]. The horizontal displacement is 5 ± 2 μm, and the loading time is 124 ± 4 ms.

The stiffness modulus is determined using Formula (8), and the Poisson ratio is determined from Formula (9):

$$S_m = \frac{F \cdot (\nu + 0.27)}{z \cdot h} \quad (8)$$

$$\nu = 3.59 \cdot \frac{z}{\Delta V} - 0.27 \quad (9)$$

where S_m—stiffness modulus of the specimen [MPa]; F—maximum force applied to the specimen [N]; ν—temperature-dependent Poisson ratio; z—amplitude of horizontal displacement of the specimen under load [mm]; h—thickness of the specimen [mm]; ΔV is the maximum vertical displacement of the specimen (corresponding to the maximum horizontal displacement) [mm].

2.4. Sample Replication

The research work was carried out according to the discussed experiment design. The laboratory tests made it possible to determine the physical and mechanical properties, which further made it possible to compare the mixtures of asphalt concrete with bitumen emulsion and indicate the impact of individual modifiers.

In order to obtain a correcC60t analysis of the properties and reliable results, six replications were performed for each ACBE mixture test. Each time, the experiment results were evaluated within a confidence interval with an assumed probability (*P* = 95%). This method of analysis made it possible to identify measurements with an error and eliminate them from further consideration. The requisite number of repetitions was determined by the necessity to stabilise the variance within the experimental design, which had been modified by the A-optimal algorithm, in conjunction with standard methodological requirements. The confidence interval was calculated according to Formula (10):

$$\overline{x} \pm 1.96\left(\frac{\sigma}{\sqrt{n}}\right) \quad (10)$$

where $\overline{x}$—expected value, 1.96—statistic for significance level α = 0.05, σ—standard deviation, n—sample size.

Planning an experiment is an extremely complex task. The common method of testing "one variable at a time" requires a significant amount of time and financial resources [59,60]. The overriding aim of planning an experiment is to obtain an answer to the question posed in an efficient manner, which means that the selection of an appropriate experiment design becomes a crucial task. Choosing the correct plan at the very beginning saves plenty of time and also money.

3. Research Results and Optimisation

3.1. Analysis of Research Results

In accordance with the adopted research design, the authors assessed the influence of modifiers on the properties of the mixture of asphalt concrete with bitumen emulsion (ACxBE). The assessment was carried out in accordance with the adopted research design, which made it possible to build mathematical models describing the analysed characteristics.

The first analysed parameter was the air void content (Vm). The obtained research results are shown in Figure 3. In the diagram, the mixtures are ranked according to the experiment design. The air void content (Vm) of the analysed mixtures of asphalt concrete with bitumen emulsion ranges between 9.8% and 18.8%.

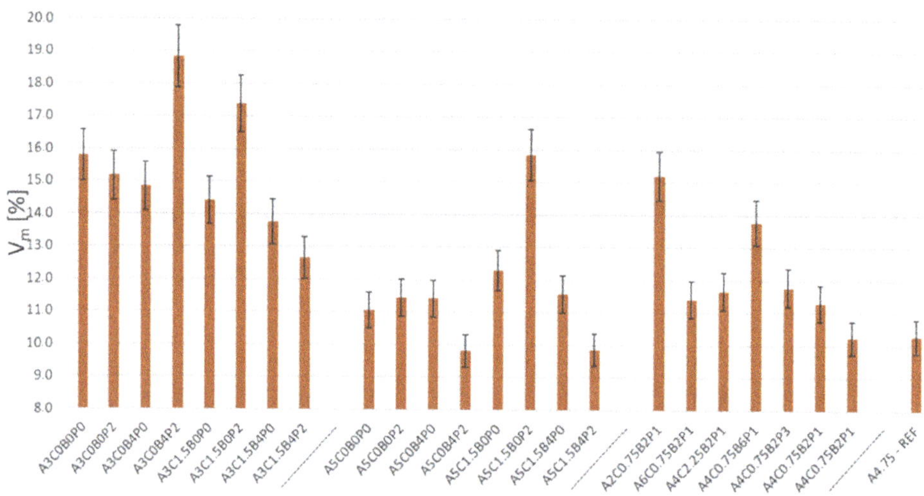

Figure 3. Air void content V_m in the analysed mixtures.

It is worth noting that the use of modifiers has a significant impact on the value of the analysed parameter. This is observable in the case of the A5C1.5B4P0 mix, which has an air void content of 11.6%. After adding 2% of redispersible polymer powder, the air void content is reduced. The mix became more workable, resulting in an air void content of 9.9%. What is more, the addition of cement also has a beneficial effect on reducing the air void content. The A3C0B0P0 mix has an air void content of 15.8%, and with the addition of 1.5% of cement (A3C1.5B0P0 mix), the air void content decreases by approximately 1.5% to 14.4%. The average air void content of the mixtures is 13.1%. The A5C0B4P2 mix has the lowest air void content, and its Vm is 9.8%. For the mixture with the highest air void content, A3C0B4P2, this parameter is 18.8%. An air void content of 10.3% was recorded for the A4.75-REF reference mix. The effect of the analysed factors on the considered Vm parameter in the ACBE mixture is shown in Table 9.

Table 9 (as well as Tables 10–13) presents the results of a response-surface regression analysis. For each property (V_m, n_w, ITS_{dry}, ITSR, S_m), a second-order polynomial model was fitted in terms of the coded contents of bitumen (A), cement (C), synthetic latex (B), and redispersible polymer powder (P). The tables list the regression coefficients ("Factor"), their t-statistics and *p*-values, together with the intercept, the coefficient of determination R^2, and the mean square of pure error (Pure error MS). Linear [(L)], quadratic [(Q)] and two-factor interaction terms (e.g., 1L × 2L) are reported. In table notation, "1L×2L" means interaction between the linear effect of factor 1 (A) and the linear effect of factor 2 (C). Finally, it describes the interaction between bitumen and cement additive.

Table 9. Effect of the analysed factors using RSM on the considered V_m parameter.

	V_m; R^2 = 0.81; Pure Error; MS = 0.6058		
	t	Factor	$p \leq 0.05$
Intercept	16.26	27.90	<0.0001
(1)A[%](L)	−7.44	−6.21	<0.0001
A[%](Q)	4.92	0.50	<0.0001
(2)C[%](L)	−2.39	−1.61	0.020
C[%](Q)	−1.23	−0.27	0.224
(3)B[%](L)	3.86	0.97	<0.0001
B[%](Q)	5.32	0.16	<0.0001
(4)P[%](L)	−2.19	−1.10	0.032
P[%](Q)	2.73	0.34	0.008
1L × 2L	8.04	1.03	<0.0001
1L × 3L	−4.53	−0.22	<0.0001
1L × 4L	1.03	0.10	0.307
2L × 3L	−13.82	−0.88	<0.0001
2L × 4L	4.03	0.51	<0.0001
3L × 4L	−6.03	−0.29	<0.0001

"R^2"—coefficient of determination; "MS"—mean square of the pure experimental error; "Intercept"—intercept of the regression model; Factor—estimated regression coefficient; "*t*"—*t*-statistic for testing H_0: $\beta = 0$; "$p \leq 0.05$"—*p*-value of the *t*-test.

Table 10. Impact of the analysed factors using RSM on the n_w parameter.

	n_w; R^2 = 0.77; Pure Error; MS = 0.09106		
	t	Factor	$p \leq 0.05$
Intercept	14.02	9.33	0.000
(1)A[%](L)	−5.63	−1.82	0.000
A[%](Q)	3.08	0.12	0.003
(2)C[%](L)	−9.12	−2.38	<0.0001
C[%](Q)	4.00	0.34	<0.0001
(3)B[%](L)	−14.41	−1.41	<0.0001
B[%](Q)	8.56	0.10	<0.0001
(4)P[%](L)	18.42	3.61	<0.0001
P[%](Q)	−3.14	−0.15	0.003
1L × 2L	8.14	0.41	<0.0001
1L × 3L	10.36	0.19	<0.0001
1L × 4L	−14.49	−0.54	<0.0001
2L × 3L	9.94	0.25	<0.0001
2L × 4L	−13.44	−0.66	<0.0001
3L × 4L	−10.52	−0.20	<0.0001

"R^2"—coefficient of determination; "MS"—mean square of the pure experimental error; "Intercept"—intercept of the regression model; Factor—estimated regression coefficient; "*t*"—*t*-statistic for testing H_0: $\beta = 0$; "$p \leq 0.05$"—*p*-value of the *t*-test.

Table 11. Impact of the analysed factors using RSM on the ITSdry parameter.

	ITS_{dry}; R^2 = 0.87; Pure Error; MS = 905		
	t	Factor	$p \leq 0.05$
Intercept	2.24	148.59	0.003
(1)A[%](L)	2.80	90.49	0.007
A[%](Q)	−1.13	−4.46	0.264
(2)C[%](L)	−3.82	−99.36	<0.0001
C[%](Q)	8.15	69.30	<0.0001
(3)B[%](L)	−2.88	−28.07	0.005
B[%](Q)	−0.30	−0.36	0.762
(4)P[%](L)	10.32	201.39	<0.0001
P[%](Q)	0.71	3.38	0.482
1L × 2L	−1.48	−7.36	0.143
1L × 3L	−0.75	−1.41	0.453
1L × 4L	−11.72	−43.68	<0.0001
2L × 3L	12.07	29.76	<0.0001
2L × 4L	6.73	33.21	<0.0001
3L × 4L	2.47	4.57	0.016

"R^2"—coefficient of determination; "MS"—mean square of the pure experimental error; "Intercept"—intercept of the regression model; Factor—estimated regression coefficient; "*t*"—*t*-statistic for testing H_0: $\beta = 0$; "$p \leq 0.05$"—*p*-value of the *t*-test.

Table 12. Impact of the analysed factors using RSM on the ITSR parameter.

	ITSR; R^2 = 0.90; Pure Error; MS = 1.024		
	t	Factor	$p \leq 0.05$
Intercept	−18.47	−77.29	0.034
(1)A[%](L)	20.04	40.23	0.032
A[%](Q)	−15.73	−3.87	0.040
(2)C[%](L)	36.45	63.04	0.017
C[%](Q)	−25.05	−13.57	0.025
(3)B[%](L)	12.77	8.28	0.050
B[%](Q)	−20.15	−1.54	0.032
(4)P[%](L)	−4.81	−6.24	0.130
P[%](Q)	−1.74	−0.53	0.332
1L × 2L	−8.94	−2.98	0.070
1L × 3L	−4.14	−0.52	0.151
1L × 4L	2.74	0.68	0.223
2L × 3L	−9.45	−1.56	0.067
2L × 4L	−10.16	−3.36	0.062
3L × 4L	9.37	1.16	0.068

"R^2"—coefficient of determination; "MS"—mean square of the pure experimental error; "Intercept"—intercept of the regression model; Factor—estimated regression coefficient; "t"—t-statistic for testing H_0: $\beta = 0$; "$p \leq 0.05$"—p-value of the t-test.

Table 13. Impact of the analysed factors using RSM on the S_m parameter.

	S_m; R^2 = 0.89; Pure Error; MS = 307,833.4		
	t	Factor	$p \leq 0.05$
Intercept	1.56	1899.87	0.124
(1)A[%](L)	1.52	883.42	0.132
A[%](Q)	−0.95	−66.35	0.346
(2)C[%](L)	−6.49	−3257.42	<0.0001
C[%](Q)	13.14	1984.82	<0.0001
(3)B[%](L)	−0.18	−34.46	0.855
B[%](Q)	−1.49	−31.47	0.140
(4)P[%](L)	10.36	3918.66	<0.0001
P[%](Q)	−3.01	−254.34	0.004
1L × 2L	0.40	39.58	0.687
1L × 3L	−0.41	−14.95	0.684
1L × 4L	−7.31	−535.65	<0.0001
2L × 3L	1.43	69.98	0.158
2L × 4L	1.35	131.71	0.182
3L × 4L	0.08	2.97	0.936

"R^2"—coefficient of determination; "MS"—mean square of the pure experimental error; "Intercept"—intercept of the regression model; Factor—estimated regression coefficient; "t"—t-statistic for testing H_0: $\beta = 0$; "$p \leq 0.05$"—p-value of the t-test.

Another analysed parameter is water absorption (N_w). Water absorption by weight is an important parameter of the mix due to its position in the system of road pavement courses. The mixtures used in road construction are constantly exposed to moisture as a result of, for example, rainfall or capillary rise. The results for the experimental ACBE mixtures are shown in the chart (Figure 4).

The analysed mixtures have a water absorption ranging from 2.7% to 5.1%. The reference mix had a water absorption of 3.2%. The lowest water absorption value was recorded for the A6C0.75B2P1 mixture. This is due to the highest asphalt content in the mix, which tightly seals the structure of the mixture. In comparison, the A2C0.75B2P1 mix with the lowest asphalt content has the highest water absorption of those analysed—5.1%. During the research, the authors encountered a problem with the A3C0B0P2 mixture, whose specimens did not survive the period of conditioning in water. The mixed specimens were destroyed, and it was not possible to determine their water absorption by weight. The authors also observed a beneficial effect of the combination of polymer powder and synthetic latex. Every time the two ingredients were used, this had a positive impact on reducing water absorption. Table 10 below shows the impact of the analysed factors on the n_W parameter.

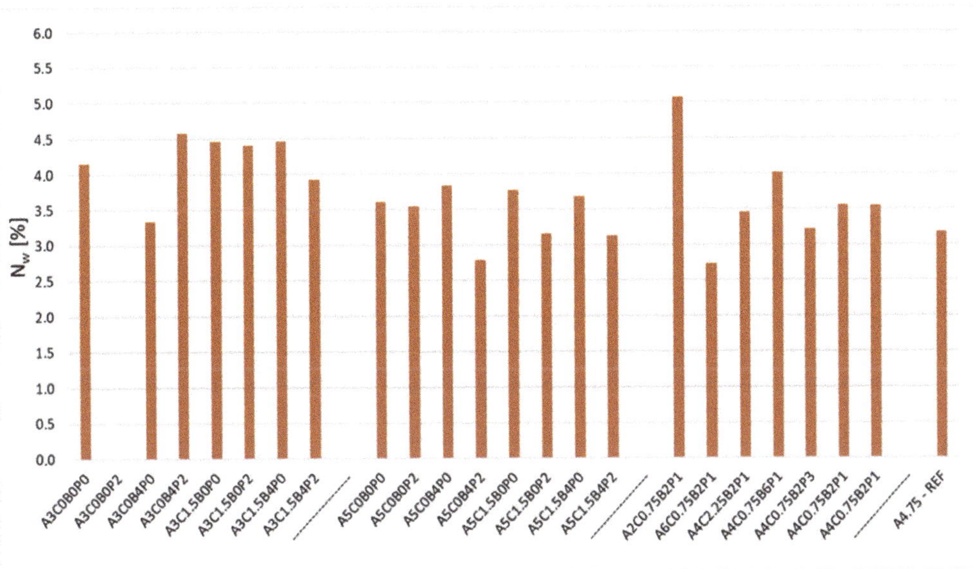

Figure 4. Air void content n_W in the analysed mixtures.

The next analysed parameter is the indirect tensile strength ITS_{DRY}. The test results for the mixtures of asphalt concrete with bitumen emulsion ranged from 222.26 kPa to 682.22 kPa. The results of the indirect tensile strength obtained from the ACBE mixtures tested in the experiment are shown in the chart (Figure 5).

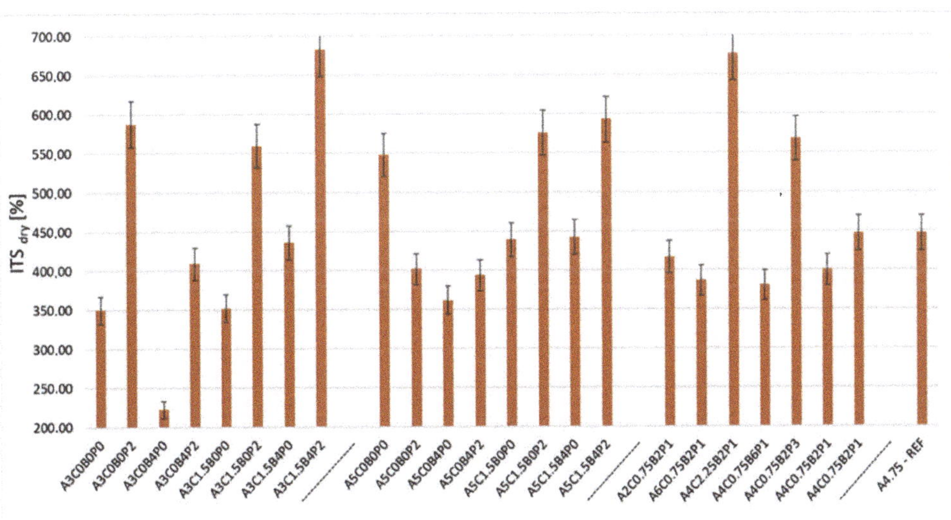

Figure 5. Indirect tensile strength ITSdry of ACBE mixtures.

The results of the ITS_{DRY} tests above show a large variation in the obtained results. This is due to the varied composition of the mixes. The achieved ITS_{DRY} value is significantly affected by the content of cement and redispersible polymer powder in the mix. This can be seen in the A5C1.5B0P2 mixtures. This mixture had an ITS_{DRY} of 575.23 kPa. In comparison, the corresponding mixture A5C1.5B0P0 had an indirect tensile strength of 438.48 kPa. The significant influence of the polymer powder is also evident in the A4C0.75B2P3 mixture. An ITS_{DRY} of 567.78 kPa was recorded for the mixture. The mixture with a reduced share of modifier P, A4C.07B2P1, had a value of 446.85 kPa. A 2% decrease in the polymer share resulted in a decrease in ITS_{DRY} by approximately 20%. The phenomenon is caused by polymer cross-linking and hydration of the cement in the mix. These mechanisms result in the stiffening of the mix. The negative impact of using synthetic latex is also noticeable. A mixture containing only 3% asphalt, designated A3C0B0P0, had an ITS_{DRY} of 349.70 kPa. The addition of 4% of latex reduced this value to 222.26 kPa. A similar phenomenon can be observed for the A5C0B0P0 mix, which recorded 547.69 kPa. Adding 4% latex to this mixture also had a negative effect, reducing the indirect tensile strength. The mixture, designated A5C0B4P0, recorded 361.63 kPa. The highest indirect tensile strength was recorded for the A3C1.5B4P2 mixture. It had an ITS_{DRY} of 682.22 kPa. The lowest value of the parameter under consideration was obtained for the A3C0B4P0 mixture—222.26 kPa. The indirect tensile strength for the A4.75-REF mix was 446.65 kPa. The observed decrease in indirect tensile strength (ITS) is caused by the lower adhesion of the asphalt precipitated from the emulsion to the aggregate grains. Adding latex to the asphalt emulsion at ambient temperatures, around 20 °C, changes the emulsion's viscosity, causing it to increase, which hinders uniform coating and inter-grain bonding. Furthermore, adding latex to the asphalt emulsion changes the surface tension and reduces the adhesion of the asphalt precipitated from the emulsion to the aggregate grains. This phenomenon is exacerbated by the fact that the ingredients are mixed "cold," at around 20 °C.

The effect of the analysed factors on the considered ITS_{DRY} parameter in the ACBE mixture is shown in Table 11.

From the perspective of the durability of ACBE mixtures, it is important to analyse frost and water damage resistance. Figure 6 below shows the results obtained from measurements of the harmful effects of water and frost on the mixture of asphalt concrete with bitumen emulsion.

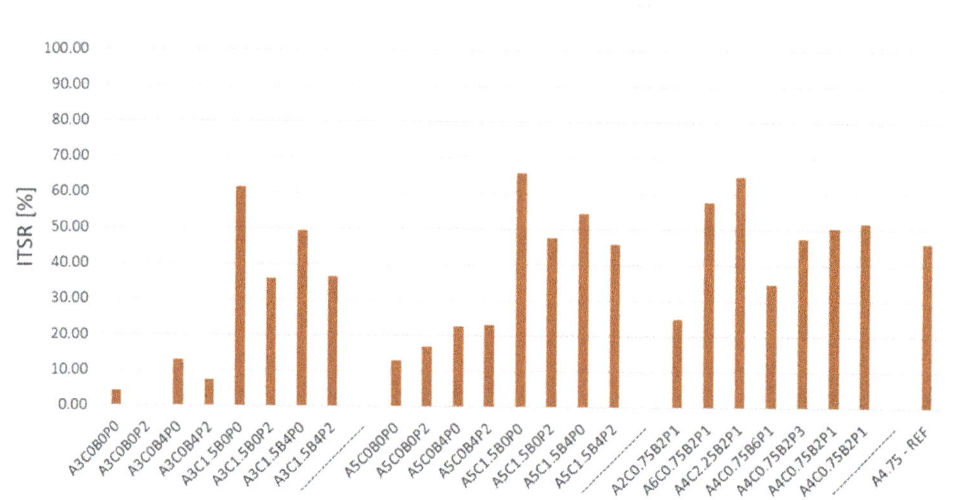

Figure 6. ITSR water and frost resistance of ACBE mixtures.

The chart above shows the recorded values for ITSR water and frost damage resistance in indirect tension. Of those analysed, only three mixtures achieved an ITSR of more than 60%. These mixtures are A3C1.5B0P0, A5C1.5B0P0, and A4C2.25B2P1. Each of the mixtures contained cement at a rate of more than 1.5%. The other modifiers made up a smaller proportion of the mix. It is therefore apparent that the use of cement has a significant effect on the parameter under analysis. In addition, this effect can be observed by comparing the mixtures containing only asphalt, i.e., A3C0B0P0 and A5C0B0P0, and juxtaposing them with corresponding mixes containing 1.5% of cement, i.e., A3C1.5B0P0 and A5C1.5B0P0. In each case, the mixtures with cement recorded values higher than 60%. A similar result was achieved by the A6C0.75B2P1 mixture, with an ITSR of 57.56%. This was significantly affected by the asphalt content of the mix, which prevented the propagation of water into the mix structure. This is apparent in the corresponding mixture containing 4% less asphalt—A2C0.75B2P1. This mixture recorded 24.90% resistance. Furthermore, the authors also recorded the negative effects of synthetic latex on ITSR. This is manifested in A4C0.75B6P1 and A4C0.75B2P1 mixes. The mixture containing 6% of synthetic latex has a water and frost damage resistance of 34.76%. Reducing the modifier content increased ITSR to 51.79%. A water and frost resistance of 46.23% was recorded for the reference mix referred to as A4.75-REF. It should also be noted that during the research, one of the mixtures did not survive until the tests. During the cycle of soaking in water, this mixture disintegrated. This indicates zero frost and water damage resistance. The worst test results were recorded for mixtures with 3% asphalt and zero cement content. These mixtures had resistance from 0.00% to 12.95%. In each case, the cement content had a positive effect on the recorded ITSR results. The effect of the analysed factors on the considered ITSR parameter in the ACBE mixture is shown in Table 12.

The last analysed parameter is the stiffness modulus of asphalt concrete with bitumen emulsion. Figure 7 below shows the results obtained from measurements. The test was conducted at 13 °C.

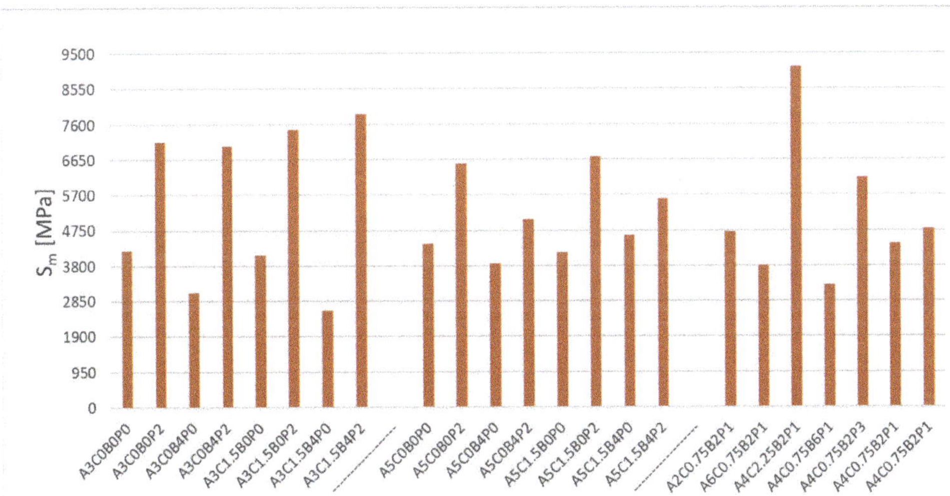

Figure 7. Value of the stiffness modulus S_m tested using the IT-CY method for ACBE mixtures.

The chart above (Figure 7) accurately illustrates the variation in the stiffness modulus depending on the types and amounts of binder used. There is a noticeable effect of the use of polymer powder on the increase in the value of the analysed parameter. The A3C0B0P2 mixture had a modulus of 7130 MPa. The lack of polymer in the A3C0B0P0 mix resulted in a stiffness modulus of 4202 MPa. Similarly, for the A5C0B0P0 mix containing 5% asphalt. The mixture achieved a stiffness modulus of 4381 MPa. With the addition of 2% of polymer powder, the stiffness modulus increased to 6533 MPa. There is also a significant impact of synthetic latex on the analysed parameter. The use of this ingredient leads to a decrease in the stiffness modulus. This is noticeable for A4C0.75B6P1 and A4C0.75B2P1 mixtures. The former of the mixtures achieved a stiffness modulus of 3275 MPa in the tests, whereas the reduction in latex share by 4% increased the modulus to 4778 MPa. Cement also played an important role. Increasing its content in the mixes leads to an increase in the stiffness modulus, as can be seen in the A4C2.25B2P1 mix, which had a stiffness modulus of 9128 MPa—the highest recorded modulus in the entire group of tested mixtures. No stiffness modulus test was carried out for the reference mix, which is why it does not appear in the chart. The effect of the analysed factors on the considered Sm parameter in the ACBE mixture is shown in Table 13.

3.2. *Optimisation*

The optimisation process involves adopting criteria whose change leads to significant changes in the context of the properties of the mixture. Accordingly, a multi-criteria optimisation of the mixture composition was carried out in order to proceed to the analyses in the second phase of the research. This method uses a general utility function [61,62] and is characterised by expression on a dimensionless scale. This type of scale requires the imposition of a range of satisfactory values. The individual criteria are described by non-negative coefficients reflecting their validity for a given mixture. The sum of the coefficients must be 1. A researcher carrying out an optimisation of this kind must be familiar with the technical requirements of the phenomenon. Using the described methodology, the authors determined the optimum quantities of the ingredients of the mixture of asphalt concrete with bitumen emulsion analysed in the project, i.e., asphalt, cement, synthetic latex, and redispersible polymer powder. The general utility function U_i^{III} contains numbers from the range (0;1). Table 14 below shows the qualitative intervals of the function.

Table 14. Assessment of the qualitative utility function [61].

Qualitative Range	Interpretation
1.00	Excellent value.
1.00–0.83	A very good value representing the achievement of material quality with extraordinary optimisation properties.
0.8–0.63	This range represents good, above-average quality.
0.63–0.37	Satisfactory (sufficient) value, acceptable under certain conditions.
0.37–0.2	Unacceptable value that may increase the unreliability of the optimised product.

The utilities attributed to the individual features y(i) are determined by two algorithms, and their profiles are shown in [49]. The general utility UIII, also denoted as D, is assumed to be the weighted geometric mean of the individual values of Formula (11):

$$D = \left[\prod_{u=1}^{n} d_u\right]^{\frac{1}{n}} \tag{11}$$

where n—number of variables

Utility function profiles illustrate two possible cases that can be encountered in the optimisation process. The optimisation task is to search for a solution range or a specific solution where the utility function adopts a value of 0.37 or higher. The optimal solution, therefore, is a result obtained in the optimisation process, taking into account predetermined criteria. Every change in the criteria significantly affects the result of the estimation of the desired outcome in terms of product properties. The work involved the optimisation of the mixture of asphalt concrete with bitumen emulsion in terms of the percentage share of asphalt, cement, synthetic latex, and redispersible polymer powder. Optimisation was carried out using the criteria shown in Table 15 below.

Table 15. Optimisation criteria adopted in the project.

	Air Voids Content (V_m)	Water and Cold Resistance (ITSR)	Stiffness (S_m), at a Tem. of 13 °C	Intermediate Tensile Strength (ITS_{dry})
	[%]	[%]	[MPa]	[kPa]
Better	1	60	5001	451
Worse	12	59	5000	450

The optimisation process used the models shown in Table 16. In the absence of existing guidelines for the design of the mixtures of asphalt concrete with bitumen emulsion, it was decided to use the guidelines for the design of mineral-asphalt mixtures (WT2—2016) [58] and the guidelines for the design of MCE mixtures [63]. Both documents are binding regulatory documents for the design and preparation of mixes in Poland. Table 16 below shows the material models for ACBE mixes.

Table 16. Material models for ACBE mixes.

Material Models	R^2
$V_m = 27.90 - 6.21{\cdot}A + 0.50{\cdot}A^2 - 1.61{\cdot}C - 0.27{\cdot}C^2 + 0.97{\cdot}B + 0.16{\cdot}B^2 - 1.10{\cdot}P + 0.34{\cdot}P^2 + 0.34{\cdot}A{\cdot}C - 0.22{\cdot}A{\cdot}B + 0.10{\cdot}A{\cdot}P - 0.88{\cdot}C{\cdot}B + 0.51{\cdot}C{\cdot}P - 0.29{\cdot}B{\cdot}P$	0.81
$ITS_{dry} = 148.59 - 90.49{\cdot}A - 4.46{\cdot}A^2 - 99.36{\cdot}C + 69.30{\cdot}C^2 - 28.07{\cdot}B - 0.36{\cdot}B^2 + 201.39{\cdot}P + 3.38{\cdot}P^2 - 7.36{\cdot}A{\cdot}C - 1.41{\cdot}A{\cdot}B - 43.68{\cdot}A{\cdot}P + 29.76{\cdot}C{\cdot}B + 33.21{\cdot}C{\cdot}P + 4.57{\cdot}B{\cdot}P$	0.87
$ITSR = -77.29 + 40.23{\cdot}A - 3.87{\cdot}A^2 + 63.04{\cdot}C - 13.57{\cdot}C^2 + 8.28{\cdot}B - 1.54{\cdot}B^2 - 6.24{\cdot}P - 0.53{\cdot}P^2 - 2.98{\cdot}A{\cdot}C - 0.52{\cdot}A{\cdot}B + 0.68{\cdot}A{\cdot}P - 1.56{\cdot}C{\cdot}B - 3.36{\cdot}C{\cdot}P + 1.16{\cdot}B{\cdot}P$	0.90
$S_m = 1899.87 + 883.42{\cdot}A - 66.35{\cdot}A^2 - 3257.42{\cdot}C + 1984.82{\cdot}C^2 - 34.46{\cdot}B - 31.47{\cdot}B^2 + 3918.66{\cdot}P - 254.34{\cdot}P^2 + 39.58{\cdot}A{\cdot}C - 14.95{\cdot}A{\cdot}B - 535.65{\cdot}A{\cdot}P + 69.98{\cdot}C{\cdot}B + 131.71{\cdot}C{\cdot}P + 2.97{\cdot}B{\cdot}P$	0.89

The optimisation resulted in eight mixtures that best match the optimisation criteria. The composition of the mixtures is included in Table 17 below.

Table 17. Composition of optimal ACBE mixtures.

Number	Symbol	A [%]	C [%]	B [%]	P [%]
1	A3.5C2B0P0	3.5	2.0	0.0	0.0
2	A3.5C2B2P0	3.5	2.0	2.0	0.0
3	A4C2B0P0	4.0	2.0	0.0	0.0
4	A4C2B2P0	4.0	2.0	2.0	0.0
5	A4.5C2B0P0	4.5	2.0	0.0	0.0
6	A4.5C2B2P0	4.5	2.0	2.0	0.0
7	A5C2B0P0	5.0	2.0	0.0	0.0
8	A5C2B2P0	5.0	2.0	2.0	0.0

A check was performed to verify the predicted models in practice. The specimens were prepared according to the predicted mixture type, and then tests under laboratory conditions were performed. The authors carried out the tests specified in the optimisation criteria. Figure 8 below shows the fit between the laboratory test results and the design parameters of the optimisation.

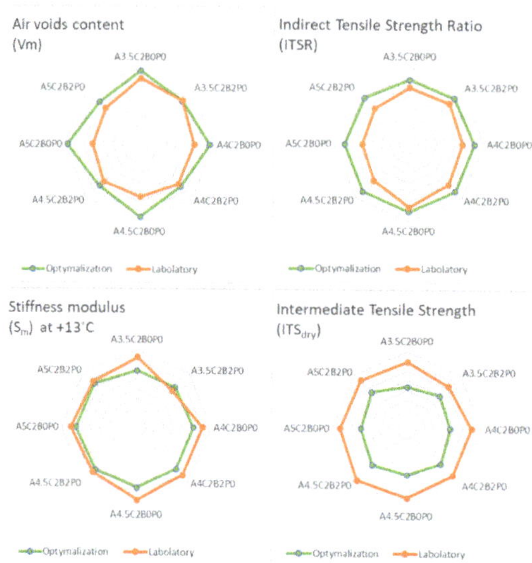

Figure 8. Fit of optimisation results.

The obtained research results indicate great potential and the possibility of using this technology as an alternative to HMA. However, it should be emphasised that the use of the ACxPB mixture should be limited to the lower layers of the structure. This requires a detailed analysis of the effect of modifiers on rheological properties and fatigue life. Recommendations for typical structural layers are also necessary after identifying the material constants used in mechanistic structural design methods. The analysed mixture offers significant environmental benefits. Production takes place at ambient temperature, without heating the aggregate. This reduces the harmful effects of environmental pollution. It also leads to financial savings by reducing the fuel consumption needed to heat the aggregate. Furthermore, the mixture can be placed without a persistent odour, which is negatively perceived by society.

4. Conclusions

The test results and analyses of the impact of the binders on the properties of ACBE asphalt concrete with bitumen emulsion support the following conclusions:

- The influence of synthetic latex is evident in the composition of the ACBE mixture. Mixtures containing modifier B are characterised by better workability, with a resultant reduction in air void content.
- The use of 1.5% cement in the mix composition has a beneficial effect on the stiffness modulus and indirect tensile strength. The use of a small amount of hydraulic binder may be necessary at the early stages of pavement construction using the ACBE mixture.
- Redispersible polymer powder adversely affects the ITSR of the mix. Specimens containing the modifier P have low frost and water damage resistance.
- The use of low asphalt content (less than 4%) and polymer powder in the mix leads to a weaker structure, compromising frost and water damage resistance.

- The use of redispersible polymer powder (P) significantly improves the workability of the mixture, as can be seen when analysing the air void content Vm.
- On the basis of the optimisation, it should be concluded that the best parameters are achieved by the C4.5C2B0P0 mixture. The other mixtures have low frost and water damage resistance.
- Rheological and fatigue life studies should be considered as future research directions.

Author Contributions: Conceptualization, M.K. (Maciej Krasowski) and P.B.; Methodology, P.B., G.M. and M.K. (Matúš Kozel); Software, G.M.; Validation, G.M. and M.K. (Matúš Kozel); Formal analysis, M.K. (Maciej Krasowski), P.B. and G.M.; Investigation, M.K. (Maciej Krasowski) and P.B.; Data curation, M.K. (Maciej Krasowski); Writing—original draft, M.K. (Maciej Krasowski); Writing—review & editing, P.B., G.M. and M.K. (Matúš Kozel); Visualization, M.K. (Maciej Krasowski) and P.B.; Supervision, P.B. and M.K. (Matúš Kozel); Project administration, P.B.; Funding acquisition, G.M. All authors have read and agreed to the published version of the manuscript.

Funding: This research was funded by Kielce University of Technology: SUBB.NSD.25.001 21.0.06.00/1.02.001.

Data Availability Statement: The original contributions presented in this study are included in the article. Further inquiries can be directed to the corresponding author.

Conflicts of Interest: The authors declare no conflict of interest.

References

1. Roberts, F.L.; Kandhal, P.S.; Brown, E.R.; Lee, D.-Y.; Kennedy, T.W. *Hot Mix Asphalt Materials, Mixture Design and Construction*, 2nd ed.; National Asphalt Pavement Association: Greenbelt, MD, USA, 1996.
2. Utracki, L.A. *Commercial Polymer Blends*; Springer US: Boston, MA, USA, 1998; ISBN 978-0-412-81020-6.
3. Błażejowski, K.; Wójcik, M.; Wiśniewska, W. *Baranowska, P. Ostrowski Poradnik Asfaltowy*; ORLEN Asfalt Sp. z o.o.: Płock, Poland, 2021.
4. Roberts, F.L.; Mohammad, L.N.; Wang, L.B. History of Hot Mix Asphalt Mixture Design in the United States. *J. Mater. Civ. Eng.* **2002**, *14*, 279–293. [CrossRef]
5. Liu, N.; Wang, Y.; Bai, Q.; Liu, Y.; Wang, P.; Xue, S.; Yu, Q.; Li, Q. Road Life-Cycle Carbon Dioxide Emissions and Emission Reduction Technologies: A Review. *J. Traffic Transp. Eng. (Engl. Ed.)* **2022**, *9*, 532–555. [CrossRef]
6. Rubio, M.D.C.; Moreno, F.; Martínez-Echevarría, M.J.; Martínez, G.; Vázquez, J.M. Comparative Analysis of Emissions from the Manufacture and Use of Hot and Half-Warm Mix Asphalt. *J. Clean. Prod.* **2013**, *41*, 1–6. [CrossRef]
7. Speight, J.G. *Asphalt Materials Science and Technology*; Elsevier: Amsterdam, The Netherlands, 2015; ISBN 978-0-12-800273-5.
8. Warm-Mix Asphalt: European Practice. 2008. Available online: https://www.asphaltinstitute.org/wp-content/uploads/The-WMA-Scan-Report-FHWA.pdf (accessed on 26 October 2025).
9. Zaumanis, M. Warm Mix Asphalt. In *Climate Change, Energy, Sustainability and Pavements*; Gopalakrishnan, K., Steyn, W.J., Harvey, J., Eds.; Green Energy and Technology; Springer: Berlin/Heidelberg, Germany, 2014; pp. 309–334, ISBN 978-3-662-44718-5.
10. Abreu, L.P.F.; Oliveira, J.R.M.; Silva, H.M.R.D.; Palha, D.; Fonseca, P.V. Suitability of Different Foamed Bitumens for Warm Mix Asphalts with Increasing Recycling Rates. *Constr. Build. Mater.* **2017**, *142*, 342–353. [CrossRef]
11. Iwański, M.; Chomicz-Kowalska, A.; Maciejewski, K. Application of Synthetic Wax for Improvement of Foamed Bitumen Parameters. *Constr. Build. Mater.* **2015**, *83*, 62–69. [CrossRef]
12. Sukhija, M.; Saboo, N. A Comprehensive Review of Warm Mix Asphalt Mixtures-Laboratory to Field. *Constr. Build. Mater.* **2021**, *274*, 121781. [CrossRef]
13. Judycki, J.; Stienss, M. Badania Laboratoryjne i Terenowe Mieszanek Mineralno-Asfaltowych o Obniżonej Temperaturze Produkcji. *Drogownictwo* **2015**, *11*, 372–379.
14. Sanchez, X.; Varamini, S. Comparison of Moisture Susceptibility of Evotherm 3G Warm Mix Asphalt versus Hot Mix Asphalt. *Can. J. Civ. Eng.* **2022**, *49*, 1885–1891. [CrossRef]
15. Stienss, M.; Judycki, J. Marcin Stienss Mieszanki Mineralno-Asfaltowe Na Ciepło—Z Asfaltem Spienionym. *Drogownictwo* **2010**, *9*, 309–312.
16. Jenkins, K.J. Mix Design Considerations for Cold and Half-Warm Bituminous Mixes with Emphasis on Foamed Bitumen. PhD Dissertation, Department of Civil Engineering, Faculty of Engineering, University of Stellenbosch, Stellenbosch, South Africa, 2000.

17. Van de Ven, M.F.C.; Jenkins, K.J.; Voskuilen, J.L.M.; Van den Beemt, R. Development of (Half-) Warm Foamed Bitumen Mixes: State of the Art. *Int. J. Pavement Eng.* **2007**, *8*, 163–175. [CrossRef]
18. Mrugała, J.; Chomicz-Kowalska, A. Influence of the Production Process on the Selected Properties of Asphalt Concrete. *Procedia Eng.* **2017**, *172*, 754–759. [CrossRef]
19. Dash, S.S.; Chandrappa, A.K.; Sahoo, U.C. Design and Performance of Cold Mix Asphalt—A Review. *Constr. Build. Mater.* **2022**, *315*, 125687. [CrossRef]
20. Jain, S.; Singh, B. Cold Mix Asphalt: An Overview. *J. Clean. Prod.* **2021**, *280*, 124378. [CrossRef]
21. Lesueur, D.; Potti, J.J. Cold Mix Design: A Rational Approach Based on the Current Understanding of the Breaking of Bituminous Emulsions. *Road Mater. Pavement Des.* **2004**, *5*, 65–87. [CrossRef]
22. Chegenizadeh, A.; Tufilli, A.; Arumdani, I.S.; Budihardjo, M.A.; Dadras, E.; Nikraz, H. Mechanical Properties of Cold Mix Asphalt (CMA) Mixed with Recycled Asphalt Pavement. *Infrastructures* **2022**, *7*, 45. [CrossRef]
23. Day, D.; Lancaster, I.M.; McKay, D. Emulsion Cold Mix Asphalt in the UK: A Decade of Site and Laboratory Experience. *J. Traffic Transp. Eng. (Engl. Ed.)* **2019**, *6*, 359–365. [CrossRef]
24. Redelius, P.; Östlund, J.-A.; Soenen, H. Field Experience of Cold Mix Asphalt during 15 Years. *Road Mater. Pavement Des.* **2016**, *17*, 223–242. [CrossRef]
25. Olard, F. LEA® (Low Energy Asphalts): A New Generation of Half-Warm Mix Asphalts. The Experience of EIFFAGE Travaux Publics. EIFFAGE Travaux Publics Research & Development Department, LEACO Technical Committee. 2007. Available online: https://proceedings-paris2007.piarc.org/ressources/files/3/IP92-Olard-E.pdf (accessed on 26 October 2025).
26. Yadykova, A.Y.; Ilyin, S.O. Nanocellulose-Stabilized Bitumen Emulsions as a Base for Preparation of Nanocomposite Asphalt Binders. *Carbohydr. Polym.* **2023**, *313*, 120896. [CrossRef]
27. Dulaimi, A.; Shanbara, H.K.; Al-Rifaie, A. The Mechanical Evaluation of Cold Asphalt Emulsion Mixtures Using a New Cementitious Material Comprising Ground-Granulated Blast-Furnace Slag and a Calcium Carbide Residue. *Constr. Build. Mater.* **2020**, *250*, 118808. [CrossRef]
28. Ling, C.; Hanz, A.; Bahia, H. Evaluating Moisture Susceptibility of Cold-Mix Asphalt. *Transp. Res. Rec. J. Transp. Res. Board* **2014**, *2446*, 60–69. [CrossRef]
29. Buczyński, P.; Krasowski, J. Optimisation and Composition of the Recycled Cold Mix with a High Content of Waste Materials. *Sustainability* **2024**, *16*, 9624. [CrossRef]
30. *Mieszanki Mineralno-Asfaltowe: Wymagania. Cz. 31: Beton Asfaltowy z Emulsją Asfaltową*; Polski Komitet Normalizacyjny: Warszawa, Poland, 2020; ISBN 978-83-8204-435-5.
31. *EN 933-1*; Tests for Geometrical Properties of Aggregates. Determination of Particle Size Distribution. Sieving Method. European Committee for Standardization: Brussels, Belgium, 2012.
32. *EN 1097-6*; Tests for Mechanical and Physical Properties of Aggregates. Determination of Particle Density and Water Absorption. European Committee for Standardization: Brussels, Belgium, 2022.
33. *EN 933-3*; Tests for Geometrical Properties of Aggregates. Determination of Particle Shape. Flakiness Index. European Committee for Standardization: Brussels, Belgium, 2012.
34. *EN 933-5*; Tests for Geometrical Properties of Aggregates. Determination of Percentage of Crushed and Broken Surfaces in Coarse Aggregate Particles. European Committee for Standardization: Brussels, Belgium, 2022.
35. *EN 1367-1*; Tests for Thermal and Weathering Properties of Aggregates. Determination of Resistance to Freezing and Thawing. European Committee for Standardization: Brussels, Belgium, 2007.
36. *EN 1097-2*; Tests for Mechanical and Physical Properties of Aggregates. Methods for the Determination of Resistance to Fragmentation. European Committee for Standardization: Brussels, Belgium, 2020.
37. *EN 1097-1*; Tests for Mechanical and Physical Properties of Aggregates. Determination of the Resistance to Wear (Micro-Deval). European Committee for Standardization: Brussels, Belgium, 2023.
38. *EN 13808*; Bitumen and Bituminous Binders—Framework for Specifying Cationic Bituminous Emulsions. European Committee for Standardization: Brussels, Belgium, 2013.
39. Krasowski, J.; Iwański, M.; Buczyński, P. Analysis of the Impact of Redispersible Polymer Powder on the Water and Frost Resistance of Cold-Recycled Mixture with Bitumen Emulsion. *IOP Conf. Ser. Mater. Sci. Eng.* **2021**, *1203*, 022006. [CrossRef]
40. *EN 197-1*; Cement. Composition, Specifications and Conformity Criteria for Common Cements. European Committee for Standardization: Brussels, Belgium, 2012.
41. *EN 196-3*; Methods of Testing Cement. Determination of Setting Times and Soundness. European Committee for Standardization: Brussels, Belgium, 2016.
42. *EN 196-1*; Methods of Testing Cement. Determination of Strength. European Committee for Standardization: Brussels, Belgium, 2016.
43. *EN 196-6*; Methods of Testing Cement. Determination of Fineness. European Committee for Standardization: Brussels, Belgium, 2018.

44. Łukowski, P. *Stowarzyszenie Producentów Cementu Modyfikacja Materiałowa Betonu*; Stowarzyszenie Producentów Cementu: Kraków, Poland, 2016; ISBN 978-83-61331-22-3.
45. Łukowski, P. *Material Modification in Concrete*; Association of Concrete Producers: Cracow, Poland, 2016; ISBN 978-83-61331-22-3. (In Polish)
46. Buczyński, P.; Iwański, M. The Influence of a Polymer Powder on the Properties of a Cold-Recycled Mixture with Foamed Bitumen. *Materials* **2019**, *12*, 4244. [CrossRef]
47. Buczynski, P.; Iwanski, M. Rheological Properties of Mineral-Cement Mix with Foamed Bitumen with the Addition of Redispersible Polymer Powder. *IOP Conf. Ser. Mater. Sci. Eng.* **2019**, *471*, 032013. [CrossRef]
48. Krasowski, J.; Buczyński, P.; Iwański, M. The Effect of Polymer Powder on the Cracking of the Subbase Layer Composed of Cold Recycled Bitumen Emulsion Mixtures. *Materials* **2021**, *14*, 5867. [CrossRef]
49. *PN-EN 12697-8:2005*; Mieszanki Mineralno-Asfaltowe—Metody Badań Mieszanek Mineralno-Asfaltowych Na Gorąco—Część 8: Oznaczanie Zawartości Wolnej Przestrzeni. European Committee for Standardization: Brussels, Belgium, 2005.
50. *PN-S-04001:1967*; Drogi Samochodowe. Metody Badań Mas Mineralno-Bitumicznych i Nawierzchni Bitumicznych. Polski Komitet Normalizacyjny (PKN): Warszawa, Poland, 1967.
51. *PN-EN 12697-23*; Mieszanki Mineralno-Asfaltowe—Metody Badań—Część 23: Oznaczanie Wytrzymałości Mieszanki Mineralno-Asfaltowej Na Rozciąganie Pośrednie. European Committee for Standardization: Brussels, Belgium, 2023.
52. *EN 12697-26*; Bituminous Mixtures. Test Methods. Stiffness European Committee for Standardization: Brussels, Belgium, 2018.
53. Box, G.E.P.; Draper, N.R. *Empirical Model-Building and Response Surfaces*; Wiley Series in Probability and Mathematical Statistics; Wiley: New York, NY, USA, 1987; ISBN 978-0-471-81033-9.
54. Cook, R.D.; Nachtsheim, C.J. A Comparison of Algorithms for Constructing Exact D-Optimal Designs. *Technometrics* **1980**, *22*, 315–324. [CrossRef]
55. Piłat, J.; Radziszewski, P. *Nawierzchnie Asfaltowe: Podręcznik Akademicki*; Wydawnictwa Komunikacji i Łączności: Warszawa, Poland, 2010; ISBN 978-83-206-1759-7.
56. Buczyński, P.; Iwański, M.; Mazurek, G.; Krasowski, J.; Krasowski, M. Effects of Portland Cement and Polymer Powder on the Properties of Cement-Bound Road Base Mixtures. *Materials* **2020**, *13*, 4253. [CrossRef] [PubMed]
57. *PN-S-04001/12*; Drogi Samochodowe i Lotniskowe. Mieszanki Mineralno-Bitumiczne. Badania. Oznaczenie Nasiąkliwości. Polski Komitet Normalizacyjny: Warsawa, Poland, 2012.
58. WT-2 Nawierzchnie Asfaltowe Na Drogach Krajowych. Mieszanki Mineralno-Asfaltowe; 2014. Available online: https://www.archiwum.gddkia.gov.pl/userfiles/articles/z/zarzadzenia-generalnego-dyrektor_13901/zalacznik%20do%20zarz%2047.pdf (accessed on 26 October 2025).
59. Kukiełka, L. *Podstawy Badań Inżynierskich*; Wydawnictwo Naukowe PWN: Warszawa, Poland, 2002; ISBN 83-01-13749-5.
60. Piasta, Z.; Lenarcik, A. Applications of Statistical Multi-Criteria Optimisation in Design of Concretes. In *Optimization Methods for Material Design of Cement-Based Composites*; E & FN Spon: London, UK; New York, NY, USA, 1998; pp. 45–59.
61. Lazić, Ž.R. *Design of Experiments in Chemical Engineering: A Practical Guide*; Wiley-VCH: Weinheim, Germany, 2004; ISBN 3-527-31142-4.
62. Montgomery, D.C. *Design and Analysis of Experiments*, 8th ed.; John Wiley & Sons, Inc.: Hoboken, NJ, USA, 2013; ISBN 978-1-118-14692-7.
63. Bogdan Dołżycki Instrukcja Projektowania i Wbudowania Mieszanek Mineralno-Cementowo-Emulsyjnych (MCE) 2019. Available online: https://www.gov.pl/attachment/53271a14-f9a8-41ca-a6e2-68c264d2f202 (accessed on 26 October 2025).

MDPI AG
Grosspeteranlage 5
4052 Basel
Switzerland
Tel.: +41 61 683 77 34

Materials Editorial Office
E-mail: materials@mdpi.com
www.mdpi.com/journal/materials

www.ingramcontent.com/pod-product-compliance
Lightning Source LLC
Chambersburg PA
CBHW062036220926
51850CB00010B/2564
* 9 7 8 3 7 2 5 8 7 3 9 4 4 *